T0130652

# Securing the U.S. Defense Information Infrastructure: A Proposed Approach

Robert H. Anderson
Phillip M. Feldman
Scott Gerwehr
Brian Houghton
Richard Mesic
John D. Pinder
Jeff Rothenberg
James Chiesa

Prepared for the
Office of the Secretary of Defense
National Security Agency
Defense Advanced Research Projects Agency

**National Defense Research Institute**

RAND

This report addresses the survivability and assured availability of essential U.S. information infrastructures, especially when they are under various forms of "information warfare" attack. To the best of our knowledge, the term "minimum essential information infrastructure" (MEII) was coined by one of the authors (Mesic) as part of the planning for a series of "Day After . . . in Cyberspace" information warfare exercises conducted from 1995 to the present under the direction of our RAND colleague Roger Molander. The idea is that some information infrastructures are so essential that they should be given special attention, perhaps in the form of special hardening, redundancy, rapid recovery, or other protection or recovery mechanisms.

Players in the "Day After . . ." exercises were intrigued by the MEII concept but asked: Is this concept feasible? Is it practical? For what portions of the Department of Defense and U.S. infrastructure is the concept relevant? What would such infrastructures look like? How effective or useful would they be? This report documents the findings of the first year of a study of the MEII concept, attempting to formulate some initial answers to these questions—or, if these are not the right questions, to ask and answer better ones. This report should be of interest to persons responsible for assuring the reliability and availability of essential information systems throughout the U.S. defense establishment, the U.S. critical infrastructure, and other organizations. Its findings and recommendations are relevant at all organizational levels, from small units to major commands.

This study is sponsored by the National Security Agency, the Defense Advanced Research Projects Agency, and the Office of the Assistant Secretary of Defense (Command, Control, Communications, and Intelligence). It is being conducted in the Acquisition and Technology Policy Center of RAND's National Defense Research Institute, a federally funded research and development center (FFRDC) sponsored by the Office of the Secretary of Defense, the Joint Staff, the unified commands, and the defense agencies. Please direct any comments on this report to the project leader,

> Robert H. Anderson
> (310) 393-0411 x7597
> Robert_Anderson@rand.org

or to the director of the Acquisition and Technology Policy Center,

> Eugene C. Gritton
> (310) 393-0411 x7010
> Gene_Gritton@rand.org

# CONTENTS

# FIGURES

# TABLES

It is widely believed, and increasingly documented, that the United States is vulnerable to various types of "information warfare" (IW) attacks. Threats range from "nuisance" attacks by hackers to those potentially putting national security at risk. The latter might include attacks on essential U.S. information systems in a major regional crisis or theater war. The purpose might be to deter (or coerce) a U.S. intervention, to degrade U.S. power projection capabilities, to punish the United States or its allies, or to undermine the support of the American public for the conflict. Critical command-and-control and intelligence systems are designed to be robust and secure under attack. However, their survivability cannot be taken for granted, and they depend on a diverse—primarily civilian and commercial—information infrastructure (consisting of the Internet and the public telephone network, among other elements).

As the diversity and potential seriousness of threats to the U.S. information infrastructure have become apparent, national-security planners and analysts have begun to think of ways to counter such threats—to increase the infrastructure's security. One immediately attractive alternative was to designate some portion of the infrastructure as the essential minimum and to harden that portion against attacks. A variant on that concept is to construct a survivable system serving the essential minimum functions, but through other means, such as dynamically reconfiguring after an attack to use remaining available resources.

In this report, we rethink the concept of a minimum essential information infrastructure (MEII) in light of the characteristics of the national information infrastructure and the nature of the threat. We suggest it is more useful to think of the MEII as a process rather than a hardened stand-alone structure, and we provide a methodology and a tool to support the implementation of that process by military units and other organizations.

## THE MEII CONCEPT

A feasible MEII may be most accurately imagined through the following contrasts:

- It does not guarantee security but is instead a type of information system insurance policy by which risks are managed at some reasonable cost while pursuing information age opportunities.

- It is not a central system responding to multiple threats but a set of systems defined locally to respond to local vulnerabilities.

- It is not a fixed, protected thing, but more likely a virtual functionality "riding on top of" the existing infrastructure.

- It is not a static structure but a dynamic process—a means to protect something, instead of a thing that has to be protected.

## SPECIFYING THE PROCESS

We propose the following six steps by which a military unit or other organizational element may implement the process approach discussed above. As more and more units implement the process, an MEII will evolve.

1. Determine what information functions are essential to successful execution of the unit's missions.

2. Determine which information "systems" are essential to accomplish those functions. Here, we use "systems" in the broadest sense, to include manual, organizational, and operational techniques.

3. For each essential system and its components, identify vulnerabilities to expected threats. In analyzing the system, it could (and perhaps should) be viewed in various ways: as a hierarchical set of

subsystems supporting each other at different levels, or as a collection of functional elements like databases, software modules, hardware modules, etc. To facilitate this step, we have developed a list of generic sources of vulnerability that can be scanned to see which apply to the system of concern (see Table S.1).

4. Identify security techniques that can mitigate each vulnerability. We have developed a list of security techniques (see Table S.2) and a matrix tool that encapsulates our abstract analysis of the applicability of various security techniques to each of the generic vulnerabilities (see Figure S.1). It also identifies unwanted byproducts of security technique application in the form of additional vulnerabilities incurred. Each protection technique, represented by a cell of the matrix in Figure S.1, is classified as follows (in response to each attribute):

— Addresses vulnerability directly and substantially (solid green).

— Addresses vulnerability indirectly or in modest degree (light green).

— Not applicable to vulnerability (blank).

— May incur vulnerability indirectly or in modest degree (light yellow).

— May incur vulnerability directly and substantially (solid yellow).

If vulnerabilities are outside the unit's control, they should be reported to higher authority, and alternative or redundant information services should be sought for backup.

5. Implement the selected security techniques.

6. Play the solutions against a set of threat scenarios to see if the solutions are robust against likely threats. It is critical that the success of security enhancements be testable. Units must have the courage to disable essential information system components during realistic exercises.

**Table S.1**

**Categories of System Vulnerabilities**

| Vulnerability | A system or process: |
|---|---|
| Inherent design/architecture | |
| Uniqueness | That is unique and may be less likely to have been thoroughly tested and perfected |
| Singularity | Representing a single point of failure, or even acting as a "lightning rod" for attacks |
| Centralization | In which all decisions, data, and control must pass through a central node or process |
| Separability | That is easily isolated from the rest of the system |
| Homogeneity | In which a flaw may be widely replicated in multiple, identical instances |
| Behavioral complexity | |
| Sensitivity | That is especially sensitive to variations in user input or abnormal use—an attribute that can be exploited |
| Predictability | Having external behavior that is predictable; attackers can know the results their actions will have |
| Adaptability and manipulation | |
| Rigidity | That cannot easily be changed in response to an attack, or made to adapt automatically under attack |
| Malleability | That is easily modifiable |
| Gullibility | That is easy to fool |
| Operation/configuration | |
| Capacity limits | Near capacity limits that may be vulnerable to denial-of-service attacks |
| Lack of recoverability | Requiring inordinate time or effort to recover operation, relative to requirements |
| Lack of self-awareness | That is unable to monitor its own use |
| Difficulty of management | That is difficult to configure and maintain, so known flaws may not be found or fixed |
| Complacency and co-optability | With poor administrative procedures, insufficient screening of operators, etc. |
| Indirect/nonphysical exposure | |
| Electronic accessibility | For which remote access provides an attack opening |
| Transparency | That allows an attacker to gain information about it |
| Direct/physical exposure | |
| Physical accessibility | In which attackers can get close enough to system components to do physical damage |
| Electromagnetic susceptibility | In which attackers can get close enough to use radiated energy to disable a system |
| Supporting facilities/infrastructures | |
| Dependency | That depends on information feeds, power, etc. |

## Table S.2

## Categories of Protection/Detection/Reaction Techniques

| | |
|---|---|
| Heterogeneity | May be *functional* (multiple methods for accomplishing an end), *anatomic* (having a mix of component or platform types), and *temporal* (employing means to ensure future admixture or ongoing diversity) |
| Static resource allocation | The a priori assignment of resources preferentially, as a result of past experience and/or perceived threats, with the goal of precluding damage |
| Dynamic resource allocation | According some assets or activities greater importance as a threat develops; this technique calls for directed, real-time adaptation to adverse conditions |
| Redundancy | Maintaining a depth of spare components or duplicated information to replace damaged or compromised assets |
| Resilience and robustness | Sheer toughness; remaining serviceable while under attack, while defending, and/or when damaged |
| Rapid recovery reconstitution | Quickly assessing and repairing damaged or degraded components, communications, and transportation routes |
| Deception | Artifice aimed at inducing enemy behaviors that may be exploited |
| Segmentation, decentralization, and quarantine | Distributing assets to facilitate independent defense and repair; containing damage locally and preventing propagation of the damaging vector |
| Immunologic identification | Ability to discriminate between self and nonself; partial matching algorithms (flexible detection); memory and learning; continuous and ubiquitous function |
| Self-organization and collective behavior | Valuable defensive properties emerging from a collection of autonomous agents interacting in a distributed fashion |
| Personnel management | Personnel security clearances and training, design of human interfaces to reduce vulnerability of systems to human frailties |
| Centralized management of information resources | Self-explanatory |
| Threat/warning response structure | Establishment of a hierarchy of increasing information attack threat levels and concomitant protective measures to be taken |

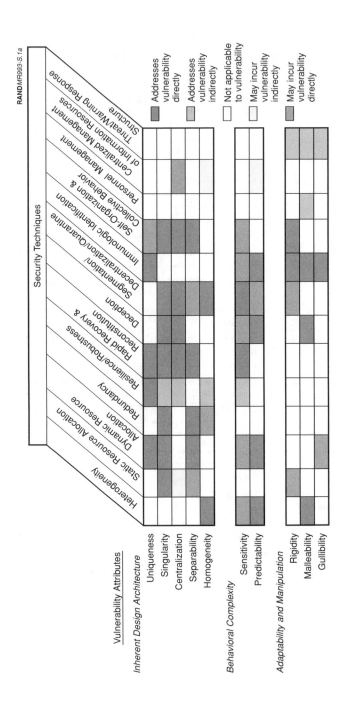

**Figure S.1—A Matrix Showing the Applicability of Security Techniques to Sources of Vulnerability**

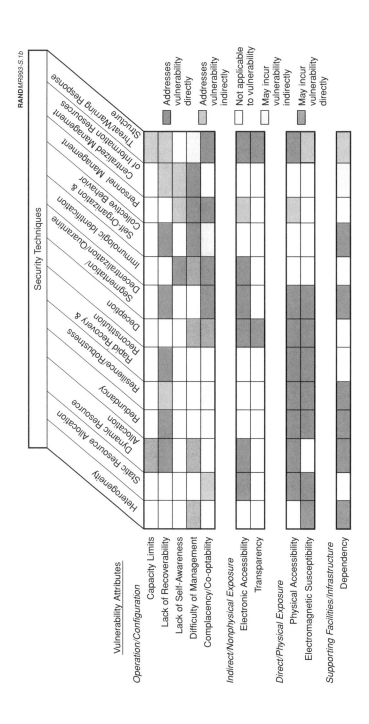

**Figure S.1—continued**

## RECOMMENDATIONS

- *Develop a test set of scenarios involving sophisticated, well-funded information warfare attacks.* Most attacks to date have been hackers probing various systems in an uncoordinated manner. Reliance on this history as a predictor of future attacks could be dangerous; we must consider what a well-funded, determined, sophisticated adversary might accomplish, and how—and at least prepare for such a contingency as a worst case.

- *Use the methodology given above as a checklist* at various Department of Defense, government, and industry unit levels, when attempting to determine which systems are essential and to ensure their survivability. We caution, however, that we have not had the time or resources to apply our methodology thoroughly to the details of a complex information infrastructure. This leads us to our next recommendation.

- *Develop case studies of our proposed methodology* to test its utility and improve its categories and procedures.

- *Explore in more detail biological analogies for robustness and survivability.* We have been impressed with the usefulness of biological analogies in terms of their power to suggest research and development (R&D) approaches that could achieve greater system survivability.

- *Consider R&D on security techniques in our list that appear to be currently underfunded.* As part of this research, we classified projects funded through certain cutting-edge programs according to which security techniques they address. We found that the distribution of effort was skewed toward certain approaches and away from others. Prominent among those lacking attention is deception.

# ACKNOWLEDGMENTS

This study benefited greatly from numerous contacts in the Department of Defense, and within the infrastructure and research and development sectors, including Lieutenant General William Donahue, Director, Communications and Information, Headquarters USAF, and Commander, Air Force Communications and Information Center; Major General John Hawley, Commander, Air and Space Command and Control Agency, Headquarters Air Combat Command; Ron Orr, Assistant Deputy Chief of Staff (Installations and Logistics), Air Force/IL; Anthony Montemarano of the Defense Information Systems Agency; Harold Sorenson, Senior Vice President and General Manager, MITRE Corporation—and their associates and members of their staffs. We have also benefited from interactions with sponsors Teresa Lunt (Defense Advanced Research Projects Agency/Information Technology Office) and Captain Richard O'Neill (ASD/Command, Control, Communications and Intelligence/Information Operations). These and other organizations and people who assisted us, however, are not responsible for any resulting misinterpretations or errors on the part of the authors of this report.

Our RAND colleague Richard Hundley was instrumental in helping to formulate and obtain funding for this project. Substantive reviews by Steven Bankes and Glenn Buchan of RAND greatly aided in improving the clarity and presentation of the report.

# ACRONYMS AND ABBREVIATIONS

| | |
|---|---|
| ACOM | Atlantic Command |
| AIN | Advanced Intelligent Network |
| API | Application Programming Interface |
| ATM | Asynchronous transfer mode |
| ATO | Air Tasking Order |
| $C^2$ | Command and control |
| $C^3I$ | Command, control, communications, and intelligence |
| $C^4I$ | Command, control, communications, computers, and intelligence |
| $C^4ISR$ | Command, control, communications, computers, intelligence, surveillance, and reconnaissance |
| CALS | Computer-aided acquisition and logistic support |
| CASE | Computer-aided software engineering |
| CERT/CC | Computer Emergency Response Team/Coordination Center (at the Software Engineering Institute, Carnegie-Mellon University, Pittsburgh, Pa.) |
| CINC | Commander in chief (of a Unified Command) |
| COE | Common operating environment (of the DII) |
| CONUS | Continental United States |
| COTS | Commercial off the shelf |
| CPE | Customer premises equipment |
| DARPA | Defense Advanced Research Projects Agency |
| DB | Database |
| DBMS | Database management system |
| DCS | Digital cross-connect system |
| DDN | Defense Data Network |
| DEFCON | Defense (Threat) condition |
| DFS | Distributed File System, by Open Software Foundation |
| DII | Defense information infrastructure |

| | |
|---|---|
| DISA | Defense Information Systems Agency |
| DNA | Deoxyribonucleic acid |
| DoD | Department of Defense |
| DRA | Dynamic resource allocation |
| DSB | U.S. Defense Science Board |
| EAM | Emergency action message |
| ECP | Engineering change proposal |
| GCCS | Global Command and Control System |
| GCSS | Global Combat Support System |
| GI | Gastrointestinal tract |
| GPS | Global Positioning System |
| GOTS | Government off the shelf |
| HEMP | High-altitude electromagnetic pulse |
| HF | High frequency |
| ICMP | Internet control message protocol |
| II | Information infrastructure |
| INFOCON | Information (threat) condition |
| IP | Internet protocol |
| IPC | Interprocess communication |
| ISDN | Integrated services digital network |
| ISO | Information Systems Office (of DARPA) |
| ITO | Information Technology Office (of DARPA) |
| IW | Information warfare |
| IW-D | Information warfare—defensive |
| IXC | Interexchange service provider |
| LAN | Local-area network |
| LEC | Local exchange carrier |
| LEO | Low Earth orbit |
| Mbps | Megabits per second |
| MCG&I | Mapping, charting, geodesy, and imaging |
| MEECN | Minimum Essential Emergency Communication Network |
| MEII | Minimum essential information infrastructure |
| MHC | Major histocompatibility complex |
| MOOTW | Military operations other than war |
| MPOA | Multiple protocols over ATM |
| MTW | Major theater war |
| NGO | Nongovernmental organization |
| NICON | National information (threat) condition |
| NII | National information infrastructure |
| NIPRNet | Sensitive But Unclassified Internet Protocol Router Network |
| NSA | National Security Agency |
| NT | The Microsoft Windows NT operating system |

| | |
|---|---|
| OPSEC | Operations security |
| OS | Operating system |
| PACOM | Pacific Command |
| PBX | Private branch exchange |
| PCCIP | President's Commission on Critical Infrastructure Protection |
| PDR | Protect/detect/react |
| PING | Packet Internet groper |
| PSN | Public Switched Network |
| PTN | Public Telecommunications Network |
| PVC | Permanent virtual circuit |
| RAM | Random access memory |
| RNA | Ribonucleic acid |
| satcom | Satellite communication |
| S/D/Q | Segmentation, decentralization, and quarantine |
| SHADE | Shared data environment (within the COE) |
| SIOP | Single integrated operational plan |
| SIPRNet | Secure Internet Protocol Router Network |
| SIW | Strategic information warfare |
| SMDS | Switched multi-megabit data service |
| SOF | Special Operations Forces |
| SONET | Synchronous optical network |
| SOS | The traditional Morse code signal for distress |
| SRA | Static resource allocation |
| SS7 | Signaling System 7 |
| STU-III | Secure telephone unit—III |
| SVC | Switched virtual circuit |
| T1 | Telephone lines with capacity 1.544 megabits per second |
| T3 | Telephone lines with capacity 45 megabits per second |
| TCP/IP | Transmission control protocol/Internet protocol |
| TPFDD | Time-phased force and deployment data |
| UHF | Ultra high frequency |
| UNIX | An operating system used, among other places, in the COE |
| VLF | Very low frequency |
| WAN | Wide-area network |
| WWMCCS | Worldwide Military Command and Control System |

While we have made a concerted effort to ensure that this report will be easily understood by readers not versed in information technology, the occasional use of technical terms is unavoidable. We define these here rather than interrupting the flow of the text.

**Asynchronous transfer mode.** ATM is a wide-area network (WAN) protocol that transfers information using fixed-size packets called cells. The 53-byte cell used with ATM is relatively small compared, for example, to Internet packets. The small, constant cell size facilitates high-speed hardware switching. Unlike other packet switched WAN protocols, ATM can support not only data (e.g., file transfers), but also connections such as real-time audio and video that require fixed latency. ATM creates a fixed circuit between source and destination when data transfer begins. This differs from transmission control protocol/Internet protocol, in which the packets of a message can take varying routes from source to destination.

**Biconnected.** Having at least two disjoint paths (i.e., having no elements in common) between any two nodes of a network.

**Domain name system (or service) (DNS).** An Internet service that translates domain names (such as rand.org) into Internet protocol (IP) addresses (such as 123.456.701.234). Whenever an alphabetic domain name is used within the Internet, a DNS must translate the name into the corresponding IP address. If a DNS server does not have the information available to translate a domain name, it asks another server, and so on, until an IP address is returned.

**Encryption, public key.** A cryptographic system that uses two keys—a public key known to everyone and a private or secret key known only to the recipient of the message. The public key is used to encrypt messages and the private key to decode them. In a properly designed system, it is virtually impossible to deduce one key from the other. Public-key cryptography is sometimes called Diffie-Hellman encryption to honor its inventors, Whitfield Diffie and Martin Hellman. It is also called asymmetric encryption because it uses two keys instead of one.

**Encryption, symmetric.** An encryption system in which the sender and receiver of a message share a single, common key that is used to encrypt and decode the message. Symmetric-key systems are simpler and faster than public-key encryption systems; however, the two parties must somehow exchange the key in a secure way.

**Ethernet.** A data transfer protocol for local area networks developed at Xerox Corporation's Palo Alto Research Center. Ethernet originally supported data transfer rates of 10 megabits per second. The newest version supports data rates of 1 gigabit (1,000 megabits) per second.

**Firewall.** Software through which those wishing to gain access to a system or set of systems must pass; intended to restrict access by potential threats.

**Internet Protocol (IP).** A protocol specifying the format of packets and the addressing scheme by which they are sent from a source node to a destination. Most networks combine IP with a higher-level protocol called Transport Control Protocol (TCP), which establishes a virtual connection between a destination and a source. An Internet Control Message Protocol (ICMP) supports administrative-type packets containing error, control, and informational messages.

**Link encryptors.** Devices that encrypt the data passing over a communication link between two host systems, so that anyone monitoring that link cannot learn the content of the transmissions.

**Local area network.** A computer network that spans a relatively small area. Each node (individual computer) connected to a LAN is able to access other data and devices on the LAN, such as laser printers or centralized databases. The LAN permits users to communicate with each other by sending e-mail or engaging in chat sessions.

**Neural network.** A software program whose operation imitates the way a human brain works. A neural network creates connections between processing elements, which are analogous to neurons. Given a certain pattern of input, the organization and weights of the connections determine the corresponding output. Most neural networks must be trained with a large set of "training data" before they can be effective in recognizing patterns within the input data.

**PING.** An acronym for "packet Internet groper," a utility that checks the accessibility of IP addresses on the network. It sends a packet to a specified address and waits for a reply. PING is used primarily to troubleshoot Internet connections. It may also be used to probe network host computers to locate candidate machines for intrusion or attack.

**Promiscuous mode.** A mode of operation for network interface devices that allows the computer to which the device is attached to monitor all data packets passing through that device, rather than only those addressed to that computer. It is therefore possible, in promiscuous mode, for a computer to "eavesdrop" on all data passing through a network.

**Root.** The top directory in a tree-structured file system. The root directory is provided by the operating system and has a special name; for example, in UNIX systems the root directory is called "/." A user with access to the root directory typically has complete control over the resources of that system, e.g., to delete or modify files, or to add or remove users.

**Router.** A device that connects two local area networks. Routers have the ability to filter messages and forward them to different places based on various criteria. The Internet uses routers extensively to forward packets from one host to another.

**Secure Telephone Unit III encryption.** A protocol for symmetric encryption between telephone units.

**Signaling System 7 (SS7).** An out-of-band signaling protocol (i.e., one employing separate communication lines) used by telephone networks to control the operation of switches within a network.

**Switched virtual circuit (SVC).** A temporary circuit that is established and used (such as through ATM) only while data is being transmitted. By contrast, a permanent virtual circuit (PVC) remains available at all times.

**Telnet.** A terminal emulation program for TCP/IP networks such as the Internet. The Telnet program runs on your computer and connects your computer to another host computer on the network. Once a telnet session is initiated, commands you enter will be executed as if you were entering them directly on the other host computer. You initiate a Telnet session by logging onto a remote host and entering a valid username and password. Permitting Telnet access to a host computer may be a vulnerability in that it permits significant user access to host facilities simply through the provision of a valid username-password combination.

**Virtual.** Having certain important characteristics of an entity but varying in essence. For example, virtual memory allows a computer to operate as if it had more memory but is realized through means other than the provision of more physical memory space.

**Wide-area network (WAN).** A computer network covering a relatively large geographical area. Computers connected to a wide-area network are often connected through public networks, such as the telephone system. They can also be connected through leased lines or satellites. The Internet is often used as a WAN.

# Chapter One

# INTRODUCTION

## PROBLEM AND PURPOSE

It is widely believed, and increasingly documented, that the United States is vulnerable to various types of "information warfare" (IW) attacks. In some scenarios, these attacks may have strategic[1] effects. Although critical command-and-control and intelligence systems are designed to be robust and secure under attack, these and many other essential systems throughout the United States depend on a diverse—primarily civilian and commercial—infrastructure. This infrastructure includes the public switched network (i.e., the telephone network), communication satellites, electric grids, air and rail traffic control, oil and gas pipelines, the financial sector, and emergency services. It has tens or hundreds of thousands of entry or access points for outsiders, and millions of trusted insiders with the ability to do damage. Concern regarding the vulnerability of the U.S. infrastructure led to the creation of the President's Commission on Critical Infrastructure Protection (PCCIP), which recently published a final report documenting many vulnerabilities in the infrastructure and recommending steps to mitigate these problems (PCCIP, 1997). Other recent documentation includes a report of the U.S. Defense Science Board (DSB) seeking to focus efforts on reducing vulnera-

---

[1]Strategic: "of great importance to an integrated whole"; "striking at the sources of an enemy's military, economic, or political power"; "essential in war"; "designed to disorganize the enemy's internal economy and destroy morale." These definitions are taken from *Webster's New Collegiate Dictionary*, Merriam Webster, 1997, and *The Concise Oxford Dictionary*, Oxford: Oxford University Press, 1982. Cited in Molander et al., 1998.

bilities in the Department of Defense (DoD) infrastructure and those systems on which it depends (DSB, 1996).

One concept explicitly considered in the deliberations of such committees is the possibility of developing a U.S. "minimum essential information infrastructure" (MEII) that could withstand—or at least degrade gracefully under—all forms of damage and IW attack.[2] When first formulated, the MEII concept was deliberately patterned after that of the Minimum Essential Emergency Communication Network (MEECN) established by the United States during the cold war. That network was designed to deliver to U.S. nuclear forces emergency action messages executing the U.S. response plan in the event of nuclear attack by the Soviet Union (for more on the MEECN, see Appendix A).

The MEECN was developed as an independent system capable of surviving a national emergency. Is that really an appropriate analogy for the MEII? In this report, we rethink the MEII concept in light of the characteristics of the national information infrastructure and the nature of the threat. We suggest it is more useful to think of the MEII as a process rather than a concept, and we devote the bulk of this report to analyses supporting the implementation of that process by military units and other organizations.

## FOCUS

The national information infrastructure is a highly complex web of systems of very different types. In its report, the PCCIP identified five infrastructure sectors separate from those dedicated to national defense (PCCIP, 1997):

---

[2]See, for example, section 6.8, "Establish and Maintain a Minimum Essential Information Infrastructure," in DSB, 1996. To the best of our knowledge, the phrase "minimum essential information infrastructure" was coined in 1995 by one of us (Richard Mesic), as part of preparations for a series of "Day After . . . in Cyberspace" exercises conducted by a RAND group headed by Roger Molander. For more on the exercises, see Molander, Riddile, and Wilson (1996); the exercises have been widely attended by senior U.S. Defense, intelligence, and essential infrastructure officials over the past several years.

- *Information and communications*—the public switched network, the Internet, and millions of computers in home, commercial, academic, and government use.

- *Physical distribution*—highways, rail lines, ports and inland waterways, pipelines, airports and airways, mass transit, trucking companies, and delivery services that facilitate the movement of goods and people.

- *Energy*—the industries that produce and distribute electric power, oil, and natural gas.

- *Banking and finance*—banks, other financial service companies, payment systems, investment companies and mutual funds, and securities and commodities exchanges.

- *Vital human services*—water supply systems, emergency services, and other government services.

An analysis of the entire national information infrastructure may not be possible; it certainly has not been possible for us, given the time and resources available. Our sponsors are all elements of the DoD or the intelligence community.[3] Because the DoD itself operates information systems essential to national defense, we have concentrated on some such systems under development and (importantly) on the portions of the national information infrastructure on which those systems depend. We hoped that through such a limited exercise some truths would emerge as to how the MEII concept might be formulated and implemented for greatest utility.

---

[3]Study funding was approximately 45 percent from the Defense Advanced Research Projects Agency, 45 percent from the National Security Agency, and 10 percent from the Office of the Assistant Secretary of Defense for Command, Control, Computers, and Intelligence.

# Chapter Two

# THE INFORMATION WARFARE THREAT AND THE MEII RESPONSE

This chapter has three objectives:

- To describe the threat to which an MEII is intended to respond.

- To set out two views of an MEII—a hardened substructure view and a process view. We conclude that the latter is more appropriate to the threat and to the nature of the information infrastructure.

- To specify the steps of our process view.

## THE THREAT

A general sense of the threat to the national information infrastructure must undergird any useful MEII concept. Here, we expand on our brief statement of the threat at the beginning of Chapter One. System vulnerabilities to specific challenges are addressed in the next chapter.

Threats range from "nuisance" attacks to those potentially putting national security at risk. There have been countless, well-documented attacks by hackers with various motives against information systems vital to the United States in peace and war. Some of these attacks appear to be coordinated, although the appearance of coordination may result only from "scripts" played by independent hackers that call upon powerful hacking toolkits available from many sites on the Internet. Such toolkits probe in seconds for hundreds of known possible vulnerabilities within any computers found in a specified Internet domain. Many of these attacks have succeeded in exposing and exploiting information system vulnerabilities.[1]

The potential threats we are concerned with here, however, have properties that distinguish them from the more common hacker nuisance attacks. Most important, the attacker must be intent on achieving certain objectives related to limiting the will or ability of the United States to protect its interests, and not simply on making mischief with a computer. To achieve his or her objectives, the serious attacker will want to be able to predict the consequences of the information system attack fairly reliably—this ability will have implications for MEII design. (For other properties of serious IW attacks, see Appendix B.)

The attacker's objectives may cover a range of threat levels. At the high end, strategic-information-warfare attacks against the United States have been envisioned by Molander, Riddile, and Wilson (1996), Molander et al. (1998), and Arquilla (1998), among others. These involve attacks on essential U.S. information systems in a major regional crisis or theater war. The purpose may be to deter or coerce U.S. intervention, to degrade power projection capabilities, to punish the United States or its allies, or to undermine the support of the American public for the conflict.

The last point is particularly important, considering the overwhelming military superiority the United States should enjoy over potential adversaries for at least the next decade. Indeed, an MEII may be more decisive where public support for a military operation is not so solid to begin with, and even more so where the forces used are not overwhelmingly superior considering the context of the operation. Such conditions are often encountered in military operations other than war (MOOTW). In such cases, attacks aimed at the nation's will to engage in the operation, or at reducing the limited capabilities of the forces committed, might "tip the balance" against the operation's success. These MOOTW-type cases are also ones in which the adver-

---

[1]We also note in passing that there are periodic fires, floods, earthquakes, tornadoes, and other natural phenomena that also threaten portions of the U.S. national information infrastructure and DoD information infrastructure.

sary might not be a nation-state against which a variety of retributions might be launched, but a terrorist organization whose operations are highly distributed, redundant, and difficult to identify or engage.

What sorts of threats might an information warrior bring to bear against the U.S. information infrastructure? The Defense Advanced Research Projects Agency (DARPA) (1997, Appendix C) has listed 21 specific threats in five categories:

- *External passive attack:* wiretapping, emanations analysis, signals analysis, traffic analysis.

- *External active attack:* substitution or insertion, jamming, overload, spoof, malicious logic.

- *Attacks against a running system:* reverse engineering, cryptanalysis.

- *Internal attack:* scavenging, theft of service, theft of data.

- *Attacks involving access to and modification of a system:* violation of permissions, deliberate disclosure, database query analysis, false denial of origin, false denial of receipt, logic-tapping, tampering.

While potential attacks can be categorized, their level of sophistication cannot be predicted. It is clear that the *capabilities* for conducting such attacks are quite widespread within an elite hacker community, and that various individuals, groups, and nation-states have demonstrated *intent* and *opportunity* to conduct various information operations.[2] At present it cannot be known with confidence whether the infrastructure elements that might be attacked would prove vital to assuring the strategic interests of the United States. Prudence, however, dictates that some effort be committed to obtaining assurance of essential information functionality.

Such efforts face impediments related to competition for resources, among others. Let us turn our attention to one impediment related

---

[2]These include—besides recreational and institutional hackers—insiders, organized crime, industrial spies, terrorists, national intelligence agencies, and information warriors (PCCIP, 1997, Figure 4).

to the nature of the threat. There are fundamental differences among the cultures, objectives, and approaches of threatening individuals and organizations, and those of security professionals. Threats and vulnerabilities are thus not just issues of hardware and software tricks and technology, but also matters of human factors, procedures, and culture. Indeed, the most worrisome threats are smart opponents who do not play by the same rules as those they attack. It is possible to focus too much on information security techniques (encryption, firewalls, codes to detect and limit damage from intruders, etc.) and too little on human and operational issues (e.g., well-trained, motivated system administrators and better user password protection procedures). It is the *people*, not hardware and software, who are often the most exploitable weak links in information systems. The most common threats exploit vulnerabilities (often well understood) that the good guys *could* fix but do not—for reasons ranging from laziness to indifference to cost to philosophical distaste.

## THE MEII AS A STRUCTURE OR PROCESS

As the diversity and potential seriousness of threats to the U.S. information infrastructure became apparent, national-security planners and analysts began to think of ways to counter such threats—to increase the infrastructure's security. One immediately attractive alternative was to designate some portion of the infrastructure as the essential minimum and to harden that portion against attacks. A variant on that concept is to construct a more robust system serving the same function as the designated minimum.[3]

Hardening of essential systems is under way and will continue. Systems deemed essential at all levels of the U.S. and defense infrastructure are being placed behind firewalls. Use of encryption for data and communication links is increasing. Intrusion detection systems, which monitor data traffic through a host computer or on a local area network, and look for abnormal patterns of behavior or known attacks, are commercially available and becoming widely deployed. There is increasing discussion of "defense in-depth," in

---

[3]The MEECN, mentioned in Chapter One and discussed briefly in Appendix A, was a robust system serving only a unique, dedicated wartime function.

which multiple levels of such hardening and monitoring are employed to catch perpetrators penetrating the initial system defenses.

Also, certain information or communication systems are so clearly part of any "essential information infrastructure" that their protection is high on any MEII list: backbones and key switches of the major U.S. telephone networks, major power grid and pipeline signaling and controlling systems, key air traffic control sites, and so on. Should the United States merely list these essential systems, harden them, and declare success?

While such an approach is useful, we do not think it is sufficient for delineating and securing an MEII. Rather than conceiving of the MEII as some minimum essential infrastructure subset secured against attack, we prefer to regard each military or organizational unit as having some minimum essential information functionality that can be secured through an ongoing process. That process is described in the next section. Here, we discuss the reasons why we do not believe the approach described above is feasible.

First, there is no single threat, and thus no single desired response that can serve as a basis for defining "essential." "Essential" begs the question, "Essential for what?" What is essential for ensuring the deployment of a U.S. fighter wing will not be the same as what is essential for sustaining the functioning of U.S. civil society in the event of a strategic IW attack.

Second, in response to premiums placed on efficiency instead of security, information-related systems and the connections and dependencies among them have proliferated. Any attempt to mark off part of the information infrastructure as "minimum essential" quickly dissolves into the realization that just about everything must be included, e.g.,

- a local telephone exchange serving a remote air base hosting a deployable air wing

- Global Positioning System (GPS) satellites and controls

- systems generating pay for DoD personnel, especially those serving in foreign theaters of operation

- the information systems of a railroad with a critical railhead or switchyard vital to certain rapid deployments
- DoD message-routing networks
- commercial off-the-shelf software used as a key part of the defense information infrastructure's common operating environment
- pipelines carrying essential fuels
- electric power to key bases and sites.

And this list does not even include systems supporting civil commerce and quality of life—systems whose integrity many people may consider to be as much a part of national security as what may be happening where a fighter wing is to be deployed.

Third, if so much is "essential," it is not all securable. Obviously, DoD has considerable control over the portions of the infrastructure it has created and operates. DoD and other government agencies have less control over portions operated by U.S. commercial companies, and they have little or no control over telecommunications and power companies in Somalia, Bosnia, or wherever the next intervention might occur. Clearly, DoD cannot ensure the security of some of these infrastructure elements, important though they may be.

Fourth, it is not really feasible to protect completely any portion of the U.S. information infrastructure. Although comprehensive vulnerability assessments have not been conducted (and, in fact, may not be feasible), it is widely understood that no *useful* computer system or network can ever be made 100 percent secure. Virtually all systems can be compromised by a trusted insider. And virtually all must be upgraded, modified, and extended to handle new information standards and opportunities, and to fix bugs and flaws. Therefore just when one expects a system to have become secure, circumstances dictate modifications that can introduce new flaws.

This brings us to the final reason why a static, structural view will not suffice: Information system security must be a dynamic concept because system structure is dynamic. A piece of the information infrastructure cannot be ossified in place as the "minimum essential" if it is to serve evolving needs. Furthermore, systems are not only evolving, they are becoming more complex. Complexity is a blessing in one sense, because increasing diversity, redundancy, adaptability, and other such network attributes *should* tend to make networks more difficult to attack. However, the complexity may be growing faster than system defenders' ability to understand and control potential threat opportunities and consequences. Recent history suggests that in the case of cyberspace security, the offense may have significant advantages over the defense. The defense must, in effect, make sure *all* threat approaches are accounted for. The offense, however, can probe until it finds just one exploitable weak link.

All of these considerations argue against the concept of an MEII as a fixed, independent bastion of secure information services immune to IW attack.[4] On the contrary,

- If it makes no sense for the MEII to be a central system responding to multiple threats, it should be defined locally to respond to local vulnerabilities.

- If the MEII cannot guarantee security, it will be more useful to think of it as a type of information system insurance policy by which risks are managed at some reasonable cost while information age opportunities are pursued.

- If the MEII cannot be a fixed, protected thing, it might be a virtual functionality "riding on top of" what remains of the existing infrastructure.

- If the MEII cannot be a static structure, it can be a dynamic process—a means to protect something, instead of a thing that has to be protected.

All four of these characteristics come to bear in responding to the situation described above in which some information infrastructure elements are beyond U.S. government control. In that event, a program of suasion, education, regulations, etc. should be adopted to ensure some level of security. As a fallback, local commanders should be urged to develop alternatives, redundancies, and

---

[4]The infeasibility of a fixed MEII is also the conclusion of an independent National Research Council report on trust in cyberspace (see Schneider, 1998).

workarounds to the extent possible.[5] To us, it makes sense to think of the MEII more in terms of a locally defined, evolving security insurance process than as a centrally secured infrastructure-wide backbone. Let us now consider what the MEII process might look like in more detail.

## STEPS OF THE MEII PROCESS

We propose six steps (listed below; see also Figure 2.1) by which a military unit or other organizational element may implement the process approach discussed above. As more and more units implement the process, an MEII will evolve and continue to evolve. Again, as should be obvious from the following, the MEII is about both systems and operations (people) that can help the country minimize risks without violating either accepted cultural norms or legal and financial constraints. The measures of MEII effectiveness must relate to the functions served by the information systems, not just the functionality of the information systems themselves. This, of course, further complicates threat and vulnerability assessments because these analyses must involve information technologists, operators, and the organizations they support—none of whom even speak the same language when it comes to such matters.

1. Determine what information functions are essential to successful execution of the unit's missions. In Figure 2.1, five information functions (IFs) are identified and three are deemed essential.

2. Determine which information "systems" (ISs) are essential to accomplish those functions. Here, we use "systems" in the broadest sense, to include manual, organizational, and operational techniques.[6] In Figure 2.1, two of the unit's three information systems are essential because they support one or more of the three essential functions.

---

[5]DoD Directive 5160.54 (January 20, 1998) contains guidance compatible with this viewpoint.

[6]Note, however, that although there may be a manual "workaround" when an information system is unavailable, performance may well be degraded, possibly to a point of jeopardizing mission objectives.

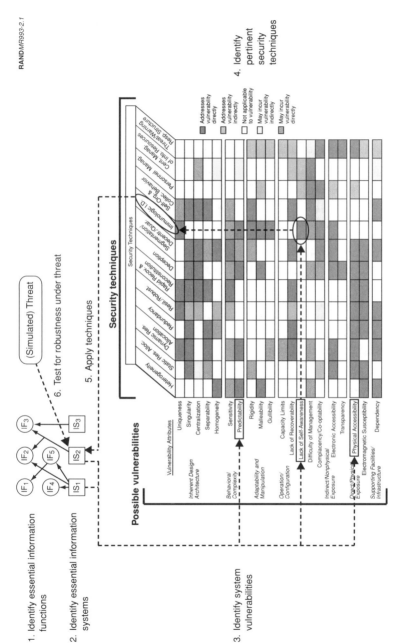

**Figure 2.1—The MEII Process**

3. For each essential system and its components, identify vulnerabilities to expected threats. In analyzing the system, it could (and perhaps should) be viewed in various ways: as a hierarchical set of subsystems in support of each other at different levels, or as a collection of functional elements like databases, software modules, hardware modules, etc. To facilitate this step, we have developed a list of generic sources of vulnerability that can be scanned to see which apply to the system of concern. In the figure, three vulnerabilities are identified as pertinent.

4. Identify security techniques that can mitigate each vulnerability. We have developed a matrix that encapsulates our abstract analysis of the applicability of various generic security techniques to each of the generic vulnerabilities. The figure shows how one of the vulnerabilities is matched with an applicable security technique by reading across the row to find a column with a green rectangle.[7] The matrix also helps ensure that the application of a particular technique does not introduce additional vulnerabilities (we will explain how later). As mentioned above, if vulnerabilities are outside the unit's control, they should be reported to higher authority, and alternative or redundant information services should be sought for backup.

5. Implement the selected security techniques. In the figure, the selected technique is applied to information system 2, which exhibits the vulnerability that the technique is capable of addressing.

6. Play the solutions against a set of threat scenarios to see if the solutions are robust against likely threats. It is critical that the success of security enhancements be testable. "Enemy" sides must have free reign in the types of information attacks they are permitted to employ during realistic exercises.

We describe the MEII process in terms of its *operational* utility. That is, given existing systems and circumstances, it provides some guidance regarding what to do to make those existing systems more ro-

---

[7]The details of the matrix will be explained in Chapter Five. Figure 2.1 is intended to convey only the existence of such a matrix and its function within the MEII process.

bust. However, elements of the process are also useful in two other important ways:

- *Developmentally:* When designing and implementing systems, designers should try to minimize the types of vulnerabilities identified in step 3 at all system levels.

- *Experimentally:* The set of security technique categories identified in step 4 points toward research and development (R&D) areas that might be valuable in increasing the survivability of systems. These solution categories should be especially valuable in providing a goal for comparison with the profile of existing research funding to determine where gaps need to be filled.

We believe this report makes two main contributions: the proposal of the six-step MEII process and the formulation of the matrix that supports steps 3 and 4. Having described the MEII process, we devote most of the rest of the report to the matrix.[8] In the next chapter, we discuss the vulnerabilities to which DoD's essential information infrastructure is subject. This can be thought of as the "demand" for security. Chapter Four turns to the "supply" of security techniques available to make essential information systems more robust—or, in the worst case, to allow them to degrade gracefully.[9] Those two chapters lay out the row stubs and column heads for the matrix. In Chapter Five, we fill in the matrix cells—which match security techniques to vulnerabilities—and make some additional observations on step 6. Chapter Six presents an analysis of the distribution of recent research that suggests areas to which more attention might profitably be paid. The body of the report ends with our conclusions and recommendations (Chapter Seven). Several topics in the text are detailed further in the appendixes.

---

[8]We will have little more to say about the steps unrelated to the matrix, because not much *can* be said about them that is not specific to a given information system, and our purpose here is to offer generally applicable guidance.

[9]Even in a degraded form, systems providing portions of an MEII may have an important role to play: to act as a communication path of last resort to reboot and reinitialize systems that have been taken down or trashed. This may require only limited bandwidth and functionality, but it is an important "minimal" feature that should be kept in mind when designing MEII-type systems at all levels.

# Chapter Three

# VULNERABILITIES

To support the implementation of the six-step MEII process outlined above, we have identified a set of generic vulnerabilities characteristic of networked information systems of the type on which the DoD and U.S. critical information infrastructures depend. In general, vulnerabilities range from straightforward physical destruction (blowing up a public switched network switching center with high explosives) to sophisticated induced failure cascades (loss of critical control networks due to cyber-attack leading to loss of service in the controlled systems, leading to further failures). Certain vulnerabilities may be exploitable only when immediate opportunities present themselves (e.g., a sniffer finds a password). These may not be exploitable in a situation in which an adversary is making war plans to be executed months in the future (the password might be changed), but they could make a mischievous hacker happy. In fact, most cyberspace vulnerabilities that have been identified (or experienced) to date are of this limited, ad hoc nature. Might there be others with more conventional military value? That is the topic of this chapter.

It will be easier to illustrate the vulnerabilities and the application of responsive security techniques if we focus our discussion on one portion of the defense information infrastructure. We thus begin with a description of a particular set of information systems. The chapter continues with some illustrations of vulnerabilities within those systems. We then present and discuss our list of generic sources of vulnerability and conclude by showing how to analyze some illustrative classes of systems for these generic vulnerabilities.[1]

## FOCUS ON EVOLVING GLOBAL COMMAND, CONTROL, COMMUNICATIONS, AND INTELLIGENCE SYSTEMS

We focus on an integrated set of systems on which many DoD operations will depend heavily in the future: The Global Command and Control System (GCCS) and the Global Combat Support System (GCSS). Focusing on these systems has at least two advantages: Our results may be applicable DoD-wide, and, since these systems are under active development, findings and recommendations resulting from our study have some chance to affect the design and deployment of aspects of these systems.

Although GCCS and GCSS have many dependencies (e.g., on the local electric power system) and linkages, we have concentrated on the net of dependencies shown in Figure 3.1. Note that Figure 3.1 does *not* represent a "wiring diagram"; the arrows only indicate dependencies between systems.[2] A dashed line separates the leased and switched portions of the public telecommunications system, since some parts of them are distinct, and other parts shared.

Of course, there are many essential systems—e.g., those supporting intelligence, surveillance, and reconnaissance—that are not shown in Figure 3.1. However, there is reason to believe that conclusions drawn from the exemplary systems shown would apply to others (in particular those with higher security classifications). For the reader unfamiliar with these systems, below we provide a brief survey of their functions and architectures.

---

[1]Further discussion of vulnerabilities within the defense information infrastructure may be found in a Defense Science Board report (DSB, 1996). See Glenn C. Buchan, "Implications of Information Vulnerabilities for Military Operations," Chapter Ten, in Khalilzad and White, forthcoming, for aspects of a recent RAND study of the vulnerabilities of U.S. Air Force information systems.

[2]For example, it is unlikely that the same instance of defense environment infrastructure–common operating environment in a workstation would be simultaneously linked with Secure Internet Protocol Router Network (SIPRNet) and the Sensitive But Unclassified Internet Protocol Router Network (NIPRNet).

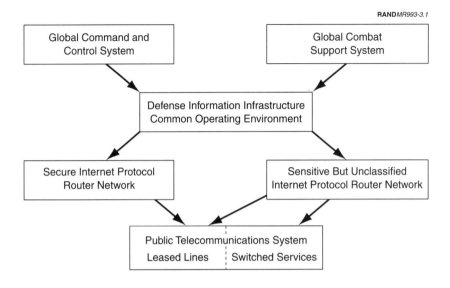

RAND*MR993-3.1*

Figure 3.1—Focus of This Study

## Global Command and Control System (GCCS)

To paraphrase the Defense Information Systems Agency's (DISA) concise description (DISA, 1995), the GCCS is specifically designed to meet the command, control, communications, and intelligence ($C^3I$) requirements of the warrior at various echelons within the command structure.  It consists of geographically distributed workstations interconnected by a classified network (the Secure Internet Protocol Router Network, or SIPRNet).  The features provided and the area network topology allow warriors to collaboratively share mission responsibilities.  Collaboration is possible in activities as diverse as creating time-phased force and deployment data (TPFDD), distributing air tasking orders, and maintaining a common view of the battlefield with up-to-date display of the deployment of all forces. The system also supports manpower requirements analysis, intelligence analysis, medical planning, office automation, and teleconferencing, among other activities.

Many of these functions are provided as separable modules. Individual commands may load onto any specific workstation only those applications that it needs, and may, with software control, restrict an individual user's access to those modules pertaining to his or her area of responsibility.

GCCS 2.0 fielding began in early 1995 at a number of sites, and version 3.0 was being installed throughout the DoD as this report was written.    GCCS serves as the near-term replacement for the Worldwide Military Command and Control System (WWMCCS).

## Global Combat Support System (GCSS)

GCSS is similar to GCCS in its distributed, networked structure, but specialized toward fulfilling acquisition and logistics support functions—e.g., workflow management and metrics, cost and schedule tracking, configuration management, and engineering-drawing support. Many of the capabilities, such as office automation, are identical to those of GCCS and hence use the same operating-environment components. Implementation of GCSS lags behind that of GCCS.

## Defense Information Infrastructure Common Operating Environment (DII-COE)[3]

The defense information infrastructure's common operating environment is a modular foundation upon which other software systems may be built, and within which they can reside. The GCCS and GCSS are among the systems built on the COE, along with various combat support, tactical, and strategic mission applications.

The COE is an open architecture designed around a client-server model. It provides a standard environment, a set of commercial off-the-shelf (COTS) components, and a set of programming standards that describes how to add new functionality to the environment. It also provides a collection of application programming interfaces for accessing COE components. It offers services such as alerts, message-processing, and data, object, and workflow management. Not

---

[3]This section is taken primarily from DISA, 1995.

all segments within the COE are required for every application. There is a "kernel COE" containing the minimal set of software required on every workstation regardless of how the workstation is used.

The COE is not tied to a specific hardware platform. As of this writing (October 1998), the three COTS operating systems supported within the COE are Windows NT, Sun Solaris (UNIX), and HP-UX (UNIX).

## Internet Protocol Router Networks

The Defense Data Network (DDN) includes two generations of networks. The older ones, such as DSNET 1, 2, and 3, are being rapidly phased out. We focus on the newer ones, which are based on the IP (Internet protocol) and on asynchronous transfer mode (ATM). These are NIPRNet (Sensitive But Unclassified Internet Protocol Router Network), and SIPRNet.

The NIPRNet is formally connected to the Internet via three gateways established for that purpose—although we have been told that many desktop workstations have links to both the NIPRNet and Internet, creating many other potential informal "gateways." The NIPRNet also shares some core network services with the Internet (in particular, domain name service). Virtually all NIPRNet traffic is unencrypted. The SIPRNet, however, was designed to be isolated from the Internet, and its connections are protected by link encryptors between the routers. Some dial-in access to SIPRNet is allowed, using STU III encryption telephone devices.

NIPRNet and SIPRNet depend on connections leased from the public switched network (PSN). However, both networks are expected to shift from leased lines to switched virtual circuits over a combination of DoD-owned and commercial ATM networks. The continental United States (CONUS) portion of the NIPRNet already makes use of commercial ATM switched virtual circuits (SVCs) to supplement the existing leased-line network.

The GCCS community relies heavily on SIPRNet for its wide-area-network infrastructure. However, the GCCS community constitutes only about 15 percent of the subscribers on the SIPRNet.

## Public Switched Network (PSN)[4]

The U.S. PSN supports analog telephony and facsimile, leased lines for voice and data, and a growing array of switched digital services, including ATM. The network is a complex of trunks, switches, database servers, digital cross-connect systems, and customer premises equipment.

It is hard to overstate the importance of the PSN. Users of the Internet, and of military networks such as the NIPRNet and SIPRNet, tend to think of those networks as though they were completely independent of the PSN. However, the networks rely on a combination of leased telephone lines and switched digital services; these in turn use the same facilities (e.g., fiber trunk lines) that support such services as analog voice telephony.

The PSN backbone consists of major switching centers (backbone offices) and high-capacity fiber trunk lines that interconnect them. The backbone is not a monolithic entity, because major service providers each have their own backbones. In a typical structure (see Figure 3.2), the country is divided into 10 parts with a class 1 switching office in each. Each of these offices can receive and send calls to the other class 1 offices and to class 2 offices (at the next level down). The remainder of the structure branches in a tree-like hierarchy down from there. A call connection request is thus sent up the hierarchy to the lowest-level switch within whose jurisdiction both the caller and the recipient reside. Since 1984, some cross-connections have been added across the tree structure, permitting lower-level call processing. These non-tree connections, which exist only in the middle and upper layers of the hierarchy, are useful for several reasons:

- They reduce the load on the higher-level switches.

- Signaling information takes fewer hops through the network, reducing call setup delay and reducing the load on both switches and trunks.

---

[4]This discussion is abstracted and updated from Feldman (unpublished).

RAND*MR993-3.2*

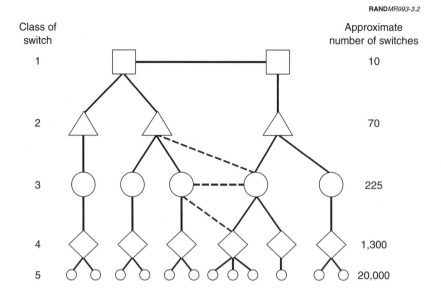

**Figure 3.2—Simplified View of Mid-1980s PSN Architecture
for the Entire United States**

• With intelligent switch control software and frequent exchanges
  of status information among switches, the network can adap-
  tively respond to congestion and to hardware failures.

The last of these is of particular importance to us: The richer mesh
topology has the potential for greater reliability and survivability
than the tree topology.

## ILLUSTRATIVE VULNERABILITIES

We now discuss some of the ways in which portions of the system
shown in Figure 3.1 are vulnerable to threats. Those portions are the
IP router networks and the public switched network.

## IP Router Networks

Since the NIPRNet and SIPRNet are based on the IP, these networks inherit most of the vulnerabilities common to other large IP-routed networks, including the following:

- *Denial-of-service attacks based on flooding.* An example of such an attack is the notorious "smurf" attack. To explain how this attack works, let $A$ denote the attacker and $V$ the victim. $A$ sends Internet control message protocol (ICMP) packet Internet groper (PING) packets containing $V$'s IP address in the sender address field, i.e., a forged return address. These forged PINGs are sent to a large number of third-party hosts, which then become unwitting accomplices in the attack when they reply to $V$. Suppose that $V$'s network is connected to the Internet (or some other larger network) through a single, fixed-rate leased line, as is commonly the case. If the total traffic volume associated with the PING responses exceeds the bandwidth of this leased line, then $V$'s network is effectively isolated by the attack. Note that a firewall at the edge of $V$'s network cannot provide protection against this type of attack.

- *Denial-of-service attacks based on forged router update messages.* Both the NIPRNet and SIPRNet use dynamic routing, which requires routers to exchange information (e.g., about link status and round-trip delays) with other routers. An attacker who sends forged router updates may be able to cause congestion that would otherwise not occur, or even cause routing loops or other anomalies.[5] The fix for router update attacks is authenticated router updates ("router hardening"). A router hardening scheme was being implemented in the SIPRNet as of late 1996 and should be completed by now.

- *Interception of unencrypted packet streams.* Since the SIPRNet uses encryption, traffic interception should be impossible, or at least impractical. For the NIPRNet, however, traffic interception is a serious concern. Within the same local-area network (LAN) as the packet source or destination, traffic can be intercepted by

---

[5]A network that uses static routing is not vulnerable to this type of attack, but is also incapable of rerouting traffic to avoid local congestion or network failures.

an Ethernet card that has been configured to operate in the so-called promiscuous mode.[6] If the attacker does not have direct access to the source or destination network, then interception is more difficult but might still be accomplished if, for instance, the attacker gains access to a router through which the packet stream of interest is likely to pass. (There is no guarantee that NIPRNet-to-NIPRNet traffic will not be routed through the Internet, although one would not expect this to happen very often.)

- *Unauthorized remote accessing of hosts.* For the SIPRNet, the concern is for insider attacks, since attacks from outside should not in principle be possible. The NIPRNet, however, has three official gateways to the Internet and, as mentioned above, is reported to have numerous improper connections to the Internet. There are no firewalls to block high-risk traffic such as telnet connections from the Internet to the NIPRNet. NIPRNet security is considered by some to be "unfixable."

- *Attacks on lower-level network protocols.* With the growing use of commercial ATM services by the DoD, ATM security is becoming increasingly important. Denial-of-service attacks are of particular concern.

- *Physical attack against the infrastructure.* Because costs are constrained, the Defense Data Network architecture emphasizes economy rather than diversity and survivability. In particular, neither the SIPRNet nor the NIPRNet has a biconnected backbone, and there is relatively little spare capacity to accommodate rerouted traffic in the event of a major failure within the network backbone.

## Public Switched Network

There is a wide spectrum of possible attacks against the PSN. It is conceptually useful to distinguish between attacks designed to deny or disrupt service and attacks intended to collect information, i.e.,

---

[6]This would not work in a LAN that uses switched Ethernet.

through traffic analysis and eavesdropping.[7] A list of selected attack types and their likely direct effects appears in Table 3.1.

Attacks designed to deny or disrupt service are potentially the most damaging and should therefore be the primary focus of attention. Disruption of telecommunications services, if carefully timed, could be useful in thwarting a U.S. response to some military action by slowing the deployment of U.S. forces.

The cross-connect systems currently operating DoD's leased lines are essentially patch boards in which connections are changed by physically unplugging and plugging in connectors. These connections are thus practically immune to manipulation or disruption unless the attacker has physical access or can disrupt electrical power.

However, to reduce leased-line costs, DISA is encouraging a transition from leased lines to switched digital services. Switched services

**Table 3.1**

**Selected PSN Attack Types and Their Likely Direct Effects**

| Attack Type | Potential Direct Effects |
|---|---|
| Sabotage end office, base "point of presence," or lines from base to end office | Disconnect a base from PSN, military networks |
| Sabotage major switches or trunks | Cause a regional PSN outage |
| Tap lines or trunk | Intercept unencrypted traffic, traffic analysis |
| Jamming of microwave radio relays | Disconnect a base from the PSN or cause a regional outage |
| Conduct dial-up or network attack against switch or digital cross-connect system | Reconfigure leased lines, cause local or regional outage |
| Conduct dial-up or network attack against private branch exchange (PBX) or integrated voice/e-mail server | Eavesdrop on conversations, fabricate voice mail |
| Implant bugs into Signaling System 7 software | Cause a regional or national PSN outage |

---

[7]Obviously, an attacker can eavesdrop on unencrypted traffic only. However, message-signaling is always unencrypted; the interception of signaling information permits one to determine where a call originated, the destination, and how long it lasted.

are handled by intelligent, stored-program devices (special-purpose computers). Some of those devices, for example, are designed to support remote administration and configuration by the service provider, as well as such caller functions as connection setup and termination, user-specified data rates, and various kinds of performance monitoring. Switched digital systems are subject to a wide range of attacks. For example, without strong authentication, one user might be able to send a command to terminate another user's connection. Even without breaching security, a user might be able to set up a connection that, at a critical time, ties up all available bandwidth on a route needed for connections to certain military bases or other DoD facilities.

Two considerations, however, should mitigate concern over the switch from leased lines to commercial switched services such as ATM. First, the duration of communication software disruption is more likely to be measured by hours than days. Historical telecommunications data (e.g., as cited in Feldman, unpublished) suggest that outages lasting longer than about eight hours typically occur only when physical damage is involved (either damage to the PSN itself or to sources of power). Attackers seeking longer disruptions may thus turn to physical attack anyway, and leased lines are also vulnerable to physical attack.[8]

Second, the public switched network is complex and redundant. It might appear that use of a system not intended to survive hostile action must make it easier for potential adversaries to effect a damaging attack. However, this would not be the case if

- DoD use of any commercial service is small relative to the total traffic load on that system

- an attacker cannot determine who is using any given piece of the system at the time of use.

An enemy seeking to disrupt DoD communications would then have to disable a large part of the commercial system in question to be

---

[8]This does not *rule out* the possibility of a lengthy disruption from a nonphysical attack.

reasonably sure of achieving his or her objective.[9]  Depending on the system architecture, this might require a coordinated attack at many points, which might exceed the resources available to the attacker. With DoD-unique networks, the smaller number of sites to be hit and the lower degree of topological diversity in the networks might make an attacker's job easier, even if individual sites and systems are better protected.[10]

Attacking commercial systems might also be undesirable for political reasons, especially if the system in question is owned or operated by an international consortium, or has users within many countries. This applies for most of the satellite trunking and telephony ventures and for all of the recent and planned undersea fiber-optic cable networks.  The system in question might even be used by businesses within the aggressor country, or by the government of that country, in which case a disruptive attack would directly harm the attacker's own interests.

## GENERIC VULNERABILITIES

In our studies of essential defense information systems, we have searched for a generic set of vulnerabilities that occur in varying degrees throughout all of them.  The purpose of this search has been to generalize beyond particular illustrative systems to problems in survivability encountered in many essential information infrastructure systems.  We have settled upon 20 generic sources of vulnerability, grouped in seven categories.[11]  These sources of possible vulnerability are listed and described below.

---

[9]However, an important exception to this statement is the tendency—and at times, mandate—to provide a "single point of presence" connecting an airfield or base to commercial networks; that connection point—usually a building near the periphery of a base—is a singularity or weakness upstream from the commercial infrastructure and the advantages we describe.

[10]We limit ourselves here to identifying possible benefits of topological diversity and other redundancy in the PSN. The trade-off between an unhardened multiplicity of sites and a hardened paucity can be illuminated by analytic resources that we cannot devote to the issue here.

[11]For other ways to classify information infrastructure vulnerabilities, see Neumann (1995) and Howard (1997).

We have tried to be as comprehensive as possible so that information system implementers and administrators in military units and other organizations can use this as a checklist to help them find important vulnerabilities in the systems they depend on. For any given system, many of the vulnerabilities listed may not apply, but we hope not to have missed too many. (We are sure to have missed some, so system administrators should not treat this comprehensive list as an exhaustive one.)

Some of the potential vulnerabilities mentioned below may seem similar to others. We separate them not simply to extend the list or to exhibit a fineness of discernment but because the differences between the vulnerabilities in question imply differences in the security techniques required to mitigate them. Nonetheless, we recognize that other analysts or system administrators may find it helpful to lump or split some of our categories—or to expand the list of vulnerabilities.

Note that each "vulnerability" may under different circumstances be a neutral or even positive attribute. For example, there is nothing inherently negative about predictability: It is often, perhaps even primarily, a desirable attribute. However, it is used here to denote only its negative aspect—i.e., that it gives attackers the advantage of knowing what results their actions will have. Many of the other vulnerabilities have a similar dual nature; some are even opposites, each member of the pair causing problems under different circumstances.

It may be of interest to note that many of these vulnerabilities derive inescapably from the technical and market constraints under which the systems exhibiting them are developed. For example, the relatively weak security models of most commercial operating systems are a direct result of market forces that have refused to accept the additional overhead and user burden of more robust models. The Multics system had a relatively robust model of this sort 30 years ago, but the perception that Multics was overburdened led to the development of UNIX, which was explicitly designed to be the opposite of Multics in many ways (even in its name). The market declared UNIX the clear winner in this contest, and Multics faded into obscurity. As will be obvious from the examples below, the pervasiveness of UNIX is not a security advantage.

Our list of 20 vulnerabilities, in seven categories, follows.

## Inherent Design/Architecture

This category includes vulnerabilities that are a direct and inherent result of the design or architecture of a system. While design or architectural decisions may give rise to other vulnerabilities that might be eliminated without major changes to the system concept, that is not the case here.

**Uniqueness.** Unique entities or processes are less likely to have been thoroughly tested and perfected.

*Example:* Legacy systems are often unique, making them vulnerable to attack.

**Singularity.** If any important component or process exists in only a single place or instance, failure of that one entity could disrupt system function and even act as a lightning rod, drawing attack as an obvious target. "Unique" and "singular" are closely linked in common parlance, but we use the former for global distinctiveness of design and the latter for in-system rarity of occurrence. By these definitions, a specially designed, one-of-a-kind component may be vulnerable by virtue of its uniqueness, regardless of how many places it occurs in a system; on the other hand, a standard, off-the-shelf component that occurs in only one place (i.e., is singular) does not suffer from uniqueness.

*Example:* Satellite communication systems often have dedicated terminals, offering a single point at which an attack could cause system failure.

**Centralization.** A system or part of a system is centralized if decisions, data, control, etc. must pass through or emanate from a single node or process. That node may also represent a singularity if it is central to the entire system. However, there may be multiple instances of subsystem centralization within a system.

*Examples:* (1) Regional or local stations that centralize switching functions may represent systemwide vulnerabilities if many can be attacked at once. (2) Some geosynchronous satellite communication systems provide user contracts for only one satellite, even if more are visible; although this does not represent a singularity for the system, any given user sees the system as centralized.

**Separability.** Components or processes that are easily isolated from the rest of a system are potentially vulnerable to attack by a "divide and conquer" strategy.

*Example:* Remote disk drives may be separable under some COE operating systems, allowing denial-of-service attacks if the operating system is cut off from these drives.

**Homogeneity.** The opposite of singularity, homogeneity refers to multiple but identical instances of a given logical entity. Although an attacker may have to attack all or many of these instances to hurt the system, the logic of the attack need only be worked out once.

*Example:* Most satellite communication systems use COTS software, databases, operating systems, etc., making them vulnerable to any single attack that disables most instances of one or more of these COTS components.

## Behavioral Complexity

Some vulnerabilities are more easily understood as manifestations of a system's behavior rather than its structure or implementation. There is not necessarily a connection between complexity (or simplicity) of behavior and structural complexity (or simplicity).

**Sensitivity.** The more sensitive a system is to variations in user input or abnormal use, the more vulnerable it may be to attack or abuse.

*Example:* The performance of an operating system (OS) may be sensitive to user load, allowing denial-of-service attacks by overloading the system, e.g., by creating multiple processes, using up file space, etc.

**Predictability.** Unlike rigidity (see below), this is an attribute of a system's behavior, not of its implementation. Wherever a system lies along the continuum between malleability and rigidity, if its external behavior is predictable, attackers have the advantage of knowing what results their actions will have, which may be a key to their success.

*Example:* The behavior of UNIX systems (widely used on the Internet) is predictable because their "internals" are widely published and known.

## Adaptability and Manipulation

This category captures those vulnerabilities that stem from the degree to which a system can be changed by direct user action or can be induced to change itself in response to such actions. While the flexibility that results from such adaptability is often of great value in a system, it may also lead to vulnerability.

**Rigidity.** Rigidity makes it harder for an attacker to modify a system maliciously, but it also implies that a system cannot easily be changed in response to an attack and cannot be made to adapt automatically under attack.

*Example:* Software that is hard-wired into the electronics of a system (e.g., within read-only memory) cannot quickly be changed to plug a security flaw that has just been discovered.

**Malleability.** Malleability is the opposite of rigidity. It is a vulnerability in that it makes it easier for an attacker to modify a system.

*Example:* Almost all software or data that reside in RAM (random access memory) or on writable disks are malleable because the bits can be changed to affect the system's operation.

**Gullibility.** This is meant to imply an attribute of a component or process that makes it inherently easy to fool. For example, systems that make no attempt to ensure that they are being used appropriately (e.g., do not check their inputs) or cannot distinguish illegitimate users or use may be vulnerable to spoofing attacks.

*Example:* Internet routers that do not check the IP address of incoming packets are gullible because they can be made to "believe" that a packet coming from an external site is really internal, and therefore to be trusted.

## Operation/Configuration

The vulnerabilities in this category stem from the ways in which a system or process is configured, operated, managed, or administered. Even if a system does not inherently suffer from these vulnerabilities, it may be used or operated in a way that forces it to exhibit one or more of them.

**Capacity Limits.** If a system's normal operating mode places it near its capacity limits, it may be vulnerable, for example, to denial of service or other problems by means of an attacker's intentionally exceeding these limits.

*Examples:* Disk space may become deliberately exhausted, forcing a system to halt operation; bandwidth on a communication channel may be fully used or allocated, keeping other messages from being sent; a telephone switch may be "war dialed" to keep all its lines in use so that no new calls can get through; Internet routers may be flooded with bad IP addresses, requiring all their processing time so that no normal routing takes place.

**Lack of Recoverability.** Since any artificial system can recover from any failure given enough time (in the extreme, by being entirely rebuilt), this vulnerability is really a measure of the speed and ease with which a system can recover. If inordinate time or effort is required to recover—relative to the needs of those who depend on the system—then it is vulnerable in this regard.

*Example:* A software system may only be "rebooted" or reinstalled through a remote telecommunications line, but that line may become unavailable as the result of an attack.

**Lack of Self-Awareness.** Systems that are unable to monitor their own use or behavior may be vulnerable to undetected attack or exploitation (depending on how accessible and sensitive they are).

*Example:* UNIX systems do not normally have the automatic capability of detecting "alien" software when it is inserted, allowing "Trojan horse" software to be installed and remain undetected, mimicking true software but having other malevolent effects.

**Difficulty of Management.** The more difficult it is to correctly configure and maintain a system, the more likely it is that an attacker can find and exploit a loophole in the configuration (for example, if known vulnerabilities are reported for a system but not fixed because of the effort required).

*Example:* Many systems (UNIX, telephone private branch exchange systems) have new periodic updates and "releases" in which many of the security defaults are reset back to "open" permissions. Managing and maintaining these systems is difficult, because increasingly elaborate retrofits of security protections must be done often and carefully. In particular, the many default settings in UNIX and its overall complexity make it difficult to configure and maintain its security features.

**Complacency/Co-optability.** Information systems are designed to assure effective human control and use. The abuse of man-machine interfaces, however, is the source of most security problems.[12] Problems can arise if personnel are lazy or incompetent. Other sources of complacency include poor administrative procedures and slack configuration control. However, the more challenging information security problem is the trusted insider who sells information or access for profit or because of ideological sympathy or because of being disgruntled and seeks revenge through misusing the systems to which he or she has access.

*Examples:* Many system installations do not have rigorous firewall configuration policies; a complacent system staff allows such known (and often published) weaknesses to be exploited. Unauthorized outsiders may exploit naïve authorized users to discover passwords that can be used to obtain unauthorized access for mischief of various sorts.

---

[12]Another leading cause of problems is simple human error, such as when a bad line of code caused massive phone outages in the Northeast. Even though they are the result of human frailty, we consider these to be technical problems best solved by technical means (e.g., good programming practices, error detection schemes, and fault-tolerant systems).

## Indirect/Nonphysical Exposure

This category consists of vulnerabilities that result from a system's being easy to access remotely (such as over a network) or being easy to study and learn about.

**Electronic Accessibility.**   Remote access to a system can be a first step toward attack.

*Example:* Just having a connection from a system to the Internet gives widespread electronic accessibility to a system, at least allowing the possibility of logical access (e.g., through password attacks) to the resources of that system.

**Transparency.**   The more open and public a system is, the easier it may be for an attacker to discover any vulnerabilities it has.

*Example:* The UNIX system and the PSN's Signaling System 7 are both "transparent" in that their codes and commands are widely published (especially within hacker bulletin boards). Once inside these systems, an intruder finds a comfortable, known environment.

## Direct/Physical Exposure

As opposed to indirect/nonphysical exposure, this category involves those vulnerabilities that allow direct, physical attack.

**Physical Accessibility.**   Physical, hands-on access to a system by more than a carefully screened, select group of operators may make it unnecessarily susceptible to attack or compromise. This may extend to communication lines, peripherals, power supplies, and other components of a system that are not obvious. Physical accessibility includes getting close enough to do physical damage, such as from a thrown grenade or a shouldered rocket launcher.

*Example:* The "single point of presence" that acts as a hub for telecommunications with a DoD base, mentioned earlier, is often on the periphery of the base and susceptible to physical damage.

**Electromagnetic Susceptibility.**   Direct access to radiated energy may make a system vulnerable to compromise, theft, intrusion, or damage.

*Example:* Almost all electronics, unless properly shielded, are susceptible to a high-energy electromagnetic pulse that would damage or disable their operation.

## Supporting Facilities/Infrastructures

Almost all information systems are dependent on exogenous supplies of power, data-input feeds, and the ability to send information to external systems. Damage, destruction, or denial of access to such supporting facilities is a source of vulnerability.

**Dependency.** The dependence of information systems on physical and other supporting infrastructures and facilities—although vital—is outside the scope of our analysis. We mention it here to remind the reader of its importance in analyzing vulnerabilities.

*Example:* Although most telephone switching centers and some essential computing centers have battery or other electric power backup, most information systems are highly dependent on continuous, reliable electric power.

## RELATIVE IMPORTANCE OF DIFFERENT VULNERABILITIES

Which of these vulnerabilities are the most important? A first step toward answering this question is to determine which ones have been involved in serious infrastructure disruptions to date. We analyzed a list of 22 of the most severe incidents reported to the Computer Engineering Response Team/Coordination Center (CERT/CC) in Pittsburgh during 1989–1995, as compiled by Howard (1997). Each of these incidents was of greater than 78 days' duration, involved more than 61 sites, and resulted in more than 86 messages to the CERT/CC. We find that all the incidents took advantage of 9 of our 20 vulnerabilities: homogeneity, predictability, malleability, gullibility, lack of self-awareness, difficulty of management, complacency/co-optability, electronic accessibility, and transparency.

On the basis of the analysis reported above, two of these vulnerabilities appear to be particularly pervasive and important: *homogeneity* and *transparency.* In the case of homogeneity, for example, we are concerned with the implications of having a standard defense infrastructure COE on tens of thousands of servers and desktops in DoD worldwide. That could lead to the replication of any security flaw (e.g., within standard releases of shrink-wrapped software, or in its installation), so that once an attack technique is discovered it might be used widely. Transparency arises most saliently from the wide availability of the source code for standard operating systems such as UNIX (used everywhere from the COE to telephone switches), and browsers such as Netscape. Similar transparency comes from the widespread posted knowledge about the Signaling System 7 codes used to control PSN switches.

This summary of experience and analysis is of some utility but should not be regarded as definitive. Indeed, it may matter less that nine vulnerabilities have been most frequently or effectively exploited than that the other eleven have yet to be. Many of these unexploited vulnerabilities could have significantly more serious consequences than those involved in previous incidents.

More generally, it is difficult to evaluate the importance of vulnerabilities because some "minor" vulnerabilities might open the door to other major ones (example: a weak password providing access to a site that is a central node in a hierarchical system). We can, however, offer a framework for thinking about importance in a practical context (see Table 3.2). In this framework, the vulnerabilities of a specific system could be placed into four classes depending on whether their damage potential is limited or serious and whether they are easy or difficult to ameliorate (or, to put it another way, whether they

Table 3.2

**Some Vulnerability Distinctions**

|  | Damage Potential | |
|---|---|---|
|  | *Limited* | *Serious* |
| Easy to fix | Type 1: easy/limited | Type 2 easy/serious |
| Hard to fix | Type 3 hard/limited | Type 4 hard/serious |

are cheap or expensive to ameliorate).[13] It is not particularly useful to assign *generic* vulnerabilities to one of the above four types. Whether a vulnerability is easy or hard to fix, or can cause limited or serious damage depends on the specific system being analyzed, and the context of that system's operation and importance.

Nonetheless, it is safe to say that most hacking, on most systems, falls into type 1. In fact, type 1 vulnerabilities probably account for the majority of real-world problems (e.g., as documented by Howard, 1997). While a nuisance, these vulnerabilities are not our concern in the MEII context. However, just because type 1 vulnerabilities ("duck bites," to quote one IW skeptic) are in the majority, that does not mean that there are no type 2 and 4 vulnerabilities to address. It means that in the absence of attention, the first indication of their importance might be an IW "Pearl Harbor" event. It also suggests that collecting and analyzing information on (detected) threat events may not be very fruitful (since most, almost by definition, will fall into type 1).

Of the two vulnerability classes with serious damage potential, type 2 vulnerabilities should be easy to address *if a military or other unit is serious about doing so.* Type 2 solutions are a matter of resolve. If the unit is serious, type 2 vulnerabilities should be and can be resolved.

Type 4, unfortunately, is not so easy to address. It is for these vulnerabilities in particular that we have created our methodology for identifying and helping to solve such problems.

## FROM GENERIC TO SPECIFIC

We now give some examples of analyzing systems to see which of the generic vulnerabilities apply and how they apply. The three illustrative types of systems analyzed are Open Software Foundation's Distributed File System (DFS), satellite communication (satcom) systems, and COE operating systems (such as Windows NT, Sun Solaris, Hewlett-Packard HP/UX). Our examples are intended to

---

[13]The two dimensions framing Table 3.2 are of course continuous variables (e.g., damage potential can take any value from negligible to catastrophic). We treat them here as categorical ("limited," "serious") to simplify the discussion.

show how to apply the methodology, not how to evaluate these system types with high confidence: Further analysis of a specific system may identify additional vulnerabilities or may reveal inaccuracies in our analysis.

Our choice of illustrative systems reflects the $C^3I$ focus described at the outset of this chapter. However, it is important to note that viewing the information infrastructure hierarchically (GCCS, COE, IP net, and so on down to switch links) is not always the most productive approach to identifying vulnerabilities. Sometimes it is better to view the infrastructure segment under consideration (e.g., GCCS and supporting systems) as a collection of databases, file systems, network services, hardware modules, and so forth. Then, these various elements can be assessed for vulnerabilities.

Our illustrative analyses are embodied in the following tables. As Tables 3.3, 3.4, and 3.5 make clear, the vulnerability of any complex system depends on how (and how well) it is configured: Analyzing a software component (such as DFS or NT) reveals its *potential* vulnerabilities, but many of these can be mitigated by conscientious administrative practices (such as plugging known loopholes, implementing available security and authentication mechanisms, etc.). The vulnerabilities of a class of hardware systems like satcom also depend on the characteristics of the specific installed system. Furthermore, vulnerabilities stemming from human failings (such as complacency or co-optability) may be largely a matter of administrative policies and attitudes, hiring practices, workload, employee satisfaction, etc. (Such organizational issues are discussed further in Chapter Six.)

These examples also suggest that our approach can be applied at several different levels of abstraction. The cases presented are all system types (such as COE operating systems), but a similar analysis can (and should) be performed on each specific system (such as Solaris or NT), as well as on specific configurations or installations of these systems. The answers to many of the questions asked by our analysis may differ at each of these levels.

The approach should also be applied at different levels of organization. Each part of a system exhibits vulnerabilities arising from its

**Table 3.3**

**Vulnerability Analysis #1,
Open Software Foundation's Distributed File System (DFS)**

| Inherent Design/Architecture | |
| --- | --- |
| Uniqueness | Not if configured properly. There are generally many copies of the system. |
| Singularity | Not if configured properly. |
| Centralization | Not if configured properly. |
| Separability | Not if configured properly. Allows replicated file systems. |
| Homogeneity | Runs on different (heterogeneous) platforms, but consists of homogeneous software. |
| **Behavioral Complexity** | |
| Sensitivity | No. Designed to be robust. |
| Predictability | Performance may not be very predictable, but behavior should be. |
| **Adaptability and Manipulation** | |
| Rigidity | No. |
| Malleability | Not generally. |
| Gullibility | Not if configured appropriately. (Authentication is available, although not always used.) |
| **Operation/Configuration** | |
| Capacity limits | May be vulnerable to flooding; may inherit OS limits, such as file size limits. |
| Lack of recoverability | Generally not bad. May be some self-healing; may require operator intervention; may use mirrored or replicated databases. |
| Lack of self-awareness | Yes. Lacks self-awareness. |
| Difficulty of management | Must be configured to avoid uniqueness and loopholes. Administration of access control lists is labor-intensive. |
| Complacency/co-optability | Varies with installation. |
| **Indirect/Nonphysical Exposure** | |
| Electronic accessibility | Not if configured appropriately. (Authentication is available but not always used.) |
| Transparency | Yes. Source code is published. |
| **Direct/Physical Exposure** | |
| Physical accessibility | Varies with physical installation. |
| Electromagnetic susceptibility | Varies with physical installation. |

## Table 3.4

## Vulnerability Analysis #2,
## Satellite Communication (satcom) Systems

| Inherent Design/Architecture | |
|---|---|
| Uniqueness | No. These systems avoid using radical, unique technology. |
| Singularity | Yes, in that they have dedicated terminals; they may have some redundancy (e.g., interoperable satellites) plus ground control (1 or 2 sites). |
| Centralization | INMARSAT, INTELSAT (GEO) provide user contracts satellite by satellite; users thus are constrained to a single satellite unless they hold multiple contracts. Teledesic/Celestri can see more than one (because of low Earth orbit (LEO)) and have a limited number of gateways to the PSN. |
| Separability | No, except by taking out gateways (a given user location may depend on a single gateway) or lines from gateway to the PSN; also some (potential) interoperability across systems would minimize this. Note: Iridium (LEO) can do satellite cross-links to use any gateway. |
| Homogeneity | Yes. The software, databases, operating systems, etc. used are generally COTS. |
| **Behavioral Complexity** | |
| Sensitivity | Not designed to be robust against hostile action (although AT&T TelStar has high power, big dish, so advertised as likely to win a power fight). Administrative databases (customer info, billing, etc.) may be vulnerable (to denial of service); generally robust against "normal accidents" but nothing else. |
| Predictability | Yes, similar to other radio communication systems. |
| **Adaptability and Manipulation** | |
| Rigidity | No active filtering, e.g., against jamming, so jamming would require *power x antenna gain* merely comparable to that of a user terminal. (Note: selective jamming would be more difficult if traffic were encrypted.) |
| Malleability | Software can be uploaded; access for this is probably no harder than for the command link. |
| Gullibility | See electronic accessibility. Little protection beyond access control. |

Table 3.4 —continued

| Operation/Configuration | |
|---|---|
| Capacity limits | Currently 90%+ utilization on existing systems. New systems (not yet running) should have extra capacity. Note: "Return-order wire" (channel-request channel, like a dial tone) has very limited capacity; carries short messages but could be flooded and is not authenticated. |
| Lack of recoverability | From physical damage: very slow (30–90 days). |
| Lack of self-awareness | Yes, except for human monitoring and onboard hardware test. |
| Difficulty of management | Not applicable, since not very configurable, although may be dynamically reconfigurable (e.g., to allocate a small number of high-capacity channels, etc.). |
| Complacency/co-optability | Only via administration of ground stations. |
| Indirect/Nonphysical Exposure | |
| Electronic accessibility | Yes. Any authentication (for user billing) is probably weak; command links are password-protected but rarely encrypted. Note: telemetry, tracking, and control are done on different terminals, using different frequencies and waveforms. |
| Transparency | Code on satellites is proprietary. Even interfaces are not always published, since terminals may be made by only 1 or 2 manufacturers. |
| Direct/Physical Exposure | |
| Physical accessibility | No, except for ground stations. |
| Electromagnetic susceptibility | Possibly vulnerable to high-power microwave effects (e.g., against LEO satellites from ground transmitter). |

own "emergent" properties, which are generally more than the sum of its parts. That is, the behavior of any specific element of the information infrastructure is not simply inherited from its components. Therefore, to ensure the robustness of the infrastructure, the vulnerabilities of every element should be analyzed, despite the fact that most elements consist of generic, ubiquitous components.

Finally, note that the approach should be applied with an open (i.e., suspicious) mind. Any system, process, component, or other infra-

## Table 3.5

### Vulnerability Analysis #3,
### COE Operating Systems (Windows NT, Sun Solaris,
### Hewlett-Packard HP/UX)

| Inherent Design/Architecture | |
|---|---|
| Uniqueness | Yes, in configuration; otherwise generally not, except for legacy OSs (e.g., in embedded systems). NT not as configurable. Legacy applications may not be able to use the COE. |
| Singularity | All have a single kernel per processor (machine). |
| Centralization | Generally not applicable. But Windows NT net administration is centralized, whereas UNIX need not be. |
| Separability | Network drives can be cut off to deny service. NFS (UNIX) does not allow replicated file systems, but DFS does (and has a better security model). NT can share file systems across networks. |
| Homogeneity | Each OS is fairly homogeneous, except for configuration (where there may be some de facto standards or guidelines). |
| **Behavioral Complexity** | |
| Sensitivity | To user actions, sensitivity is limited to denial of cycles, file space, etc. |
| Predictability | Yes, fairly high for all. |
| **Adaptability and Manipulation** | |
| Rigidity | HP/UX has no access control lists, so permissions are rigid, which may encourage operational compromises; access control lists are provided by Solaris; NT 5.0 has a hierarchical domain security model and flexible permission lists; NT has no virtual machine model. |
| Malleability | Fairly customizable and configurable, e.g., in kernel. NT less than UNIX, but multiuser protection fairly good in UNIX; NT is not sold as multiuser, although it does have user privileges (can be extended to be multiuser with 3rd-party add-ons of variable malleability). |
| Gullibility | Some, known. |

Table 3.5—continued

| Operation/Configuration | |
|---|---|
| Capacity limits | Fixed (compile-time) table sizes for maximum numbers of open files, active processes, buffer/pipe sizes, etc. NT workstation has limits, e.g., 10 users. |
| Lack of recoverability | Fairly good recovery from file system separation. Ability to boot disconnected from net (quarantine). |
| Lack of self-awareness | Yes. All have minimal self-awareness (e.g., auditing). |
| Difficulty of management | This is variable but may be getting better: All have administrative tools of varying degrees of user-friendliness. NT's access control model is powerful but complex to configure and administer. |
| Complacency/co-optability | Varies with physical installation. |
| **Indirect/Nonphysical Exposure** | |
| Electronic accessibility | Access to OS itself is prevented by passwords (if implemented!); source code is generally proprietary, but interfaces are well documented. |
| Transparency | Loopholes are publicized by system administrator alerts; source code is generally proprietary, but interfaces are well documented. UNIX configurability reduces transparency. |
| **Direct/Physical Exposure** | |
| Physical accessibility | Varies with physical installation. |
| Electromagnetic susceptibility | Varies with physical installation. |

structure element to be analyzed with this approach should be thought of as a suspect being interrogated about the vulnerabilities identified above. Each element in question should be asked *how* it exhibits each vulnerability, assuming that it probably exhibits the vulnerability in some manner. This analysis is not unlike that performed by an attacker examining a system: To make systems less vulnerable, the defender should adopt the perspective of an attacker and assume that most systems exhibit most vulnerabilities in some way and to some degree. If the defender fails to find a particular vulnerability in a system, he or she should not infer that the system is invulnerable in this respect: The defender should conclude instead that he or she has not been as probing, imaginative, or resourceful as an attacker is likely to be.

## A CONCLUDING OBSERVATION

We conclude this discussion of vulnerabilities by returning to our point about the relationship between utility and robustness that we made in reference to the transition from Multics to UNIX. We believe that ultra-efficiency and absolute dependability may be enemies of MEII-type survivability. That is, the less-dependable systems may well be the most survivable in the broadest sense. Our reasoning is as follows: Because they are less dependable, people will have developed alternatives and workarounds (forming part of the larger "system," which may include manual techniques, other communication paths, etc.) that can be relied upon if the system truly becomes unavailable. Systems that are "always" dependable (except when directly attacked by malevolent actors) do not encourage the exploration of workarounds and alternatives.

# Chapter Four

# RESPONSIVE SECURITY TECHNIQUES

To identify security techniques that might address the vulnerabilities identified in Chapter Three, we reviewed research and development efforts in information system architecture and design and in cyberspace security and safety, and drew on our own experience in these areas. It is noteworthy that some researchers are applying biological principles associated with survivability and protection to information-based systems. Some research on information system security is even "biomimetic," in the sense that it uses approaches that mimic biological systems such as the human immune system. A detailed discussion and analysis of these biological metaphors and analogies, and their application to information systems, is provided in Appendix C. See Chapter Six for more on the status of research on information system security techniques.

In this chapter, we provide a brief definition and description of categories of security techniques, each of which is intended to perform one or more of the following security functions: make the system less vulnerable in advance of an attack, detect an attack, and react to an attack once it is initiated. This chapter includes one or more examples of how techniques of each type can be applied to increase the security and survivability of an information-based system.

It is important to realize at the outset that technical constraints make it impossible or infeasible to eliminate certain vulnerabilities by straightforward redesign. Technical solutions may not yet exist, they may be too expensive to implement, or they may have undesirable consequences—they may eliminate desired performance characteristics or generate other vulnerabilities. For example, systems allowing multiple users are inherently vulnerable to sabotage, but if multiple use is a primary design criterion, eliminating it by restricting access to the system defeats its purpose.

For this reason, the security techniques suggested below cannot simply be opposites of the vulnerabilities they seek to redress. Rather, they are techniques that are intended to mitigate the effects of inherent vulnerabilities without undermining desirable design and market decisions that may have led to those vulnerabilities.

As with the generic vulnerabilities discussed in Chapter Three, we think the distinctions we make below have functional utility, but we do not presume to have come up with the only sensible taxonomy of security techniques. We should also point out here that some types of security techniques (e.g., immunologic identification) are at earlier stages of development than others, so our discussions of them are somewhat more speculative.

## HETEROGENEITY

Heterogeneity is essentially the presence of diversity, or nonuniformity, within a system. The diversity associated with heterogeneity makes it more difficult for an attack to exploit one type of flaw or weakness in many different places throughout a system. Heterogeneity may be expressed in several complementary modes and at several levels. A system might be *spatially* heterogeneous, in that different—but functionally similar—components are used at different locations within a distributed system. *Temporal* heterogeneity might be accomplished (most easily in software) by rearranging the locations of items (e.g., in a computer's memory) over time.

*Examples:* (1) Use of multiple telephone lines, from different service providers, for telecommunication to and from a military base. (2) Deliberate mixing of operating systems (e.g., Windows NT and UNIX) among the personal computers and workstations within a command. (3) Use of randomizing compilers and software linkers and loaders so that individual software routines are placed in different physical locations within the memories of disparate computers in an installation.

## STATIC RESOURCE ALLOCATION

A second class of protection techniques is the a priori preferential assignment of resources based on experience and the perceived threat environment. The elements of the system that are deemed to be the most critical, or at the greatest risk, are provided with more resources and hence greater protection. This category also includes selective hardening techniques that provide critical portions of a system with extra fixed defenses. This is not a flexible technique: The protective resources cannot be reallocated in response to short-term changes in the threat environment, and fixed defense cannot easily be dismantled in one part of a system and then reapplied somewhere else.

*Examples:* Keeping essential system software in unalterable read-only memory is an application of selective hardening to protect critical system components. Static resource allocation can be used to alleviate the physical vulnerabilities of an information system by locking up, or guarding, the facilities that house critical data and hardware components.

## DYNAMIC RESOURCE ALLOCATION

System resources may be allocated to various assets and activities based on their relative importance, given the current state of the system and any perceived threats. The system protects itself dynamically by responding quickly to perceived changes in its environment. Dynamic resource allocation most often requires prioritization of activities and use of information assets. (See Appendix D for further discussion of prioritization in information systems.)

Dynamic resource allocation also implies that each commander should have thought about which of his or her activities and assets, in various scenarios, are most important to the execution of that mission. Then, in an emergency, it will be clear which activities and assets should be given highest priority when faced with scarce bandwidth, scarce computational resources, and so on. Such planning is now under way in some commands as part of INFOCON-level, information (threat) condition, planning (see below).

*Example:* An information system can use dynamic resource allocation to protect itself by prioritizing clients and processes and by reducing nonessential functions, based on the current system loading and any observed or anticipated threat activity.

## REDUNDANCY

Multiple system components, or duplicates of key information, can be made available to replace or compensate for any portions of a system that are lost, damaged, or corrupted as a result of an accident, natural failure, or attack. The protection provided by redundancy is increased when the spares or backups are not colocated with the primary components but can still be activated in a timely manner to maintain system functionality. Note that this set of techniques is different from those mentioned under "heterogeneity," which achieves multiplicity of structure or function through diversification, while redundancy achieves it through replication.

*Example:* Mirrored databases provide an information system with protection through redundancy. By keeping multiple up-to-date versions of a database in different geographic (and topological) locations, access to essential, reliable data can be maintained even if one of the databases is corrupted, cut off, or lost.

## RESILIENCE AND ROBUSTNESS

This category of protection techniques includes anything that enables the individual components of a system to thwart or absorb an attack without the system as a whole experiencing any significant degradation in availability or performance. This is the category within which most current information system security measures fall, and in which most R&D is focused (see Chapter Six).

Two existing, important techniques for achieving robustness in a system are firewalls and use of encryption. Firewalls are computer systems that monitor transmissions and requests, only allowing authorized persons or types of data to pass through. They may be used at various levels and degrees of stringency to create a series of nested system "enclaves," with increasing requirements for access.

Encryption is perhaps the most ubiquitous and important technique for achieving robustness in an information system. Systems for transmitting classified data, such as SIPRNet, use hardware-based encryption between network nodes, thus protecting the links. Systems for electronic commerce increasingly use combinations of public–key encryption and symmetric encryption for the security of those transactions. Important debates are now under way regarding the differing means of encryption, its export beyond U.S. borders, its control and restriction by various governments, etc. For further discussion of encryption, see, for example, Schneier (1996) regarding the technology, and Dam and Lin (1996) regarding policy issues.

Within this general category of protection techniques, we distinguish between resilience and robustness as follows:

- *Resilience:* System components affected by an attack are able to resume their normal functions so quickly that the system is not noticeably affected.

  *Example:* Excess capacity—whether in terms of processor speed, transmission bandwidth, or memory—can enable an information system to be more resilient if it is attacked.

- *Robustness:* System components are sufficiently "tough" to be able to resist, or absorb, most types of attack without suffering major loss of functionality.

  *Example:* An information system that uses strong encryption for transmitting or storing critical or private data is robust, since it is extremely difficult for an attacker to decode the encrypted information, even if it can be surreptitiously obtained. Robustness can also be improved by using software and hardware that is either fault-tolerant or has been thoroughly tested and debugged, and by using software languages and techniques that aid security.

## RAPID RECOVERY AND RECONSTITUTION

A system may be provided with the capability to recover or reconfigure itself in a timely manner, so as to minimize any disruption in service or performance resulting from an attack. Such techniques often necessitate some type of damage detection and assessment capability. They generally entail the activation of solutions or procedures designed to handle a particular class of attacks or contingencies.

*Example:* Survivability can be improved by enabling a system to closely monitor its own status and performance, detect and assess a potential attack, and quickly respond by implementing a streamlined rebooting process that includes, e.g., switching to dedicated backup servers.

## DECEPTION

Well-applied deceptions have aided combatants in both offense and defense throughout history and the breadth of conflict—from insurgency to invasion. We define deception here as artifice to induce exploitable enemy behaviors; such craft has great potential worth in the protection and survivability of information assets.

The means of deception frequently take the form of camouflage, ruses, feints, decoys, and disinformation. But deception is more than just assets or practices; it is a planning process that combines operational flexibility with subterfuge to manipulate and exploit the enemy's behavior. The first objective of any deception is to affect beliefs in the mind of the enemy: altering his or her perceptions, creating confusion, diverting his or her attention, and the like. The second objective is to provoke a response that may be capitalized upon, such as a commitment or withholding of forces or a misplacement of efforts.

For information assets, successful deceptions may provide enhanced survivability and protection by avoiding attacks outright, or by hampering, or even harming, would-be adversaries. Deception may serve as the outermost hedge against attackers, because of the proactive nature of guile and entrapment. For a more detailed discussion of battlefield deception, see Appendix E.

*Example:* A system, or a portion of it, might be made attractive to attackers by posting files containing provocative keywords and data. This "honeypot" may detain perpetrators long enough to allow better understanding of their sophistication, modus operandi, and interests, and to allow traceback of their access route. Such deceptively attractive files may, in addition, steer perpetrators away from more sensitive sites and files. Files within a system might also be mislabeled to prevent identification of content. (Note: Deception techniques must be used very cautiously, to avoid disrupting a unit's operations.)

## SEGMENTATION, DECENTRALIZATION, AND QUARANTINE

Protection techniques in this category improve the security and survivability of a system by containing and isolating local damage to prevent it from spreading to other parts of the system. The three different approaches are as follows:

- *Segmentation:* Portions of a system may be enabled to function autonomously, so that it is difficult for damage incurred in one area to bring down the entire system.

  *Example:* The survivability of a large information network can be improved by dividing it up into various "islands," each of which can, if necessary, function as a self-contained independent network. (Under normal circumstances, these network islands would be integrated.)

- *Decentralization:* The critical nodes or functions of a system can be widely distributed to reduce the chance that disruption at one point would disable the entire system. Decentralization often involves the local detection and management of intrusions since this eliminates the delay and additional resources associated with a centralized response process.

  *Example:* Each node of a network can be given all the information needed to reboot and reconfigure itself or any adjacent node. The authority to make certain important, or potentially dangerous, changes in an information system could also be decentralized by requiring the simultaneous approval of multiple administrators. This would protect the system against co-opted or disgruntled insiders intent on causing harm.

- *Quarantine:* When damage or an intrusion is detected in a particular region of a system, that region is immediately isolated from the rest to prevent propagation of the damage, or movement of the attacker, or both. The regional isolation can occur in any of several dimensions, including geographical, topological, logical, and chronological.

*Example:* Information systems can use one-way gateways to isolate any portion of the system whenever an attack, or even suspicious activity or damage, is detected. This type of quarantine scheme would necessarily rely on continuously operating automated intrusion and damage detection, to locate the region affected and then to isolate it from the rest of the system.

## IMMUNOLOGIC IDENTIFICATION

The techniques in this category are modeled after biological immune systems, which distinguish and destroy intruding cells. These techniques (and those in the next category) are not all implementable at this time; some are still in concept or development stages. We include them to form as complete a list as we can of techniques likely to be available in the near future, to indicate their utility, and to encourage more research and development along these lines. There are, however, two commercially available examples of defense against computer virus attacks modeled largely on immunologic principles: Those offered by IBM (Kephart et al., 1997) and the "AutoImmune" system of Network Associates Inc. (www.nai.com).

As applied to information systems, immunologic action includes four key capabilities:

1. *Self/Nonself Discrimination:* Unfamiliar entities and anomalous behaviors could be detected and investigated to determine if they are foreign or improper.

   *Example:* Unauthorized users could be identified as such by observing their activities and behavior as well as their credentials. An alarm would be set off if, for example, a dormant user account became active at two o'clock in the morning and then acquired root privileges.

2. *Partial Matching Algorithms:* The system defenses could recognize not only known or understood threats, but also hazards that have not been encountered. This would build flexibility into sys-

tem defenses and substantially reduce the amount of threat information they need to be effective.

*Example:*  Hackers invading an information system could be identified using a few common behavior patterns, rather than a large database of individual characteristics.  In this way, new hackers could be identified as easily as old ones, provided that they behave in a manner similar to other hackers.  Modern "intrusion detection" systems placed on network hosts or LANs exhibit a form of this behavior.

3. *Memory, Adaptation, and Communication:* Information on intrusion attempts could be recorded, along with the type and effectiveness of the defensive countermeasures taken.  These data could be used to improve the defense to be mounted against similar attacks in the future, whether at the same location or not.

*Example:*  In an information system, this type of approach could be implemented by incorporating neural networks, or other pattern recognition techniques capable of learning, into intrusion detection and response schemes.  By learning and adapting, the resulting system might efficiently provide protection from frequent attacks by threats similar to each other.

4. *Continuous and Ubiquitous Defense:* The system could interrogate and examine all entities, independent of their identity or status.  There would be no "safe haven" where an invader could evade the possibility of detection.

*Example:*  Autonomous roving software agents could be used to observe and question any user anywhere in an information system.  This type of defense would prevent an attacker from "setting up camp" anywhere in the system, since there is always a chance he or she would be discovered.  To detect and intercept co-opted or disgruntled insiders, even the most trusted users could be observed and interrogated periodically.

## SELF-ORGANIZATION AND COLLECTIVE BEHAVIOR

This category includes adaptive approaches to security that rely on self-organization within a system, and the associated emergence of advantageous collective behaviors.  Many examples of these techniques have evolved in nature. For example, hives of bees or nests of ants are essentially self-organized systems in the sense that individuals undertake behavior promoting system welfare without instructions from a central authority. Characteristics of self-organizing systems include the following:

- *Goal-Oriented Behavior:* Individual system components act to optimize some objective using a set of simple rules. It is important to note that these rule sets are not closed; each individual may experiment with new rules that are either generated internally, or learned by observing or communicating with other individuals, while also discarding less-effective, older rules. Thus, individuals in a self-organized system adapt and evolve over time to improve their performance and their chances of survival.

  *Example:* Some active information and communications networks use adaptive processes, such as rule-based systems, genetic algorithms, or neural networks, to seek greater security and reliability. Such systems are self-organizing, to the extent that they are not centrally controlled.

- *Specialization:* A system organizes itself in a manner that facilitates both efficiency and security. Different types of individuals emerge that are specialized to perform certain functions or respond to certain circumstances. Some parts of the system may work to improve efficiency, while others may provide protection against infrequent but dangerous threats to the system.

  *Example:* In some sense, the use of certain computers as servers, firewalls, and terminals is a form of specialization within an information system that improves the system's security. A more apt example of self-organized specialization would be a system that assigns security-related duties, such as user checks and system diagnoses, to different computers in the network, based on current system status and threat environment. It is likely that, over time, certain parts of such a system would become specialized for specific security duties, by virtue of their characteristics and usage patterns.

## PERSONNEL MANAGEMENT

The categories of security techniques described so far entail enlisting technology in the service of security. The current category and the next two involve modifications in human behavior and institutions.

The unwitting compromise of passwords and other security breaches can be minimized through threat awareness indoctrination, training and education, monitoring, and, as appropriate, penalties for violating security procedures. People whose actions can compromise information system security can be made aware of this potential and urged to take their responsibilities to protect these systems with appropriate seriousness. They can be informed about how hostile agents might attempt to compromise information systems, so that they minimize their personal vulnerability and detect and report potentially hostile activities. Training can be made to seem more relevant through tailoring to the specific responsibilities and access of the trainees.

Deliberate security compromises might be reduced through more rigorous background checks, but, as countless well-documented espionage cases attest to, security programs can never be perfect. Thus, critical human interfaces might be designed to make human frailties less of a concern. Several techniques might help without being too much of a financial or operational burden. These fall into three categories: (1) limiting individual access, authority, and control; (2) monitoring and reporting critical activities; (3) requiring two people to concur and cooperate on system-critical functions (e.g., operating system changes).

*Examples:* Unsophisticated users might require only periodic reminders of the importance of protecting their passwords and access to their terminals; system administrators could be more thoroughly trained. In military units, information system administration might be made a well-rewarded career track, so that expertise built up through training and education could be utilized in the long term for the unit's benefit; too often, system administration is a short-term, part-time assignment for someone whose main career incentives (at least within the military) lie elsewhere.

## CENTRALIZED MANAGEMENT OF INFORMATION RESOURCES

Three of the security techniques identified above—heterogeneity, dynamic resource allocation, and redundancy—require the availability of alternative systems or elements, e.g., multiple communication channels with differing physical, electromagnetic, and protocol characteristics. If individual units, at varying levels of the command structure, "own" these communication assets, they are unlikely to be available, rapidly or automatically, to other units needing some access to them. This problem could be addressed through central management of an array of information options and assets (perhaps including some dedicated to military functions, and others leased or "rented" from commercial firms).[1]

*Example:* To some extent, the centralization of many information assets in the DISA is a trend that is compatible with this category of security techniques.

## THREAT/WARNING RESPONSE STRUCTURE

DoD has long used a set of five defense (threat) condition (DEFCON) levels to indicate the level of threat facing the United States. At each level, there are well-defined actions to be taken to afford greater protection of essential defense assets and greater readiness of troops and materiel. An analogous set of threat levels—sometimes called "INFOCON" or "NICON," national information (threat) condition, levels—could be important in alerting units and commands to the level of information warfare threat being faced. Actions could be taken at the various levels to provide additional protection or reaction against that threat.[2]

---

[1]Another option for enhancing the availability of alternatives in the field is the design and installation of antennas, protocol converters, and even connectors that allow differing information links to be rapidly accessed.

[2]In considering actions to be taken at each INFOCON level, commanders must first decide which information processes and assets are most important to their missions. If the MEII process we propose here were implemented, that prioritizing would already have been undertaken (in steps 1 and 2).

*Examples:* Several U.S. commanders in chief (e.g., the CINCs of Atlantic and Pacific Commands) have instituted such INFOCON definitions, listing explicit sets of actions to be taken at each level. Roger Molander et al. (1998), in reporting on exercises illuminating threats to the U.S. infrastructure and responses to those threats, have defined five NICON levels, giving example actions to be taken by various infrastructure sectors at each threat level. An example of the types of actions recommended in the Molander report is shown for each threat level in Table 4.1, which is adapted from that report.

## PROTECT, DETECT, REACT

As noted above, the various security techniques are intended to protect an information system by making it less vulnerable in advance of an attack, to detect an attack, or to react to an attack once one is initiated. Our security technique categories may be apportioned among these goals as follows (there is of course some overlap):

- *Protect:* Heterogeneity; static resource allocation; redundancy; resilience and robustness; deception; segmentation, decentralization, and quarantine; and personnel management.

- *Detect:* Immunologic identification.

- *React:* Dynamic resource allocation, rapid recovery and reconstitution, self-organization and collective behavior, centralized management of information resources, and threat/warning response structure.

## FROM GENERIC TO SPECIFIC

To save space, we have typically limited our examples to one for each category or subcategory of security techniques. As a result, it may not be obvious that each generic category comprises a wide variety of specific measures, the choice of which depends on the system to be secured. For example, heterogeneity applied to a radio communications link might mean using different frequency ranges, different modulation schemes, or different multiplexing and coding schemes. Heterogeneity applied to administration of a computer system might mean defining different administrative processes or sequences of actions for different installations. And of course, not all techniques can be meaningfully applied to all systems.

## Table 4.1

### Example INFOCON Threat Levels and Corresponding Actions

| | |
|---|---|
| Green | **Owners and Operators**<br>• Report Category I incidents weekly to Sector Center |
| Yellow | **Owners and Operators**<br>• Round-the-clock staffing of Sector Center<br>• Report new Category I and II incidents within 24 hrs to Sector Center<br>• Increase physical security including<br>— Enhance remote site surveillance and security<br>• Increase network security including<br>— Restrict access to Internet<br>— Increase frequency of intrusion detection reporting<br>• Place "call preemption" capability (software and personnel) in standby mode<br>• Review and update INFOCON ORANGE and INFOCON RED contingency plans |
| Orange | **Owners and Operators**<br>• Report new Category I, II, and III incidents immediately to Sector Center<br>• Further increase physical security including<br>— Physical protection of critical network elements<br>• Further increase network security including<br>— Restrict access to Internet<br>— Further restrict remote access to key system elements<br>• Notify nonessential users and database managers that they may be denied access to the network at any time<br>• Nonessential network users must follow a preapproved and preprioritized checklist to fulfill their mission requirements<br>• Increase frequency of software backups<br>• Prepare to use backup power (e.g., alert diesel vendors to possible demand)<br>**State/Federal Assistance**<br>• Deploy selected National Guard and Reserve units to<br>— Augment owner/operator security at critical nodes<br>— Defend and assist in reconstitution of PSN and emergency communications<br>— Ensure telecommunications support for possible military contingencies |

## Table 4.1—continued

| Red | Owners and Operators |
|-----|---------------------|
| | • Implement "call preemption" capability |
| | • Remove nonessential users from unclassified networks (access to only those preprioritized critical users) |
| | • Place critical database and information on classified networks or at least remove the sensitive database from unclassified networks and servers |
| | • Implement workarounds to get nonessential information to customers via other means |
| | • Ensure disaster recovery capabilities are staffed and operable |
| | **State/Federal Assistance** |
| | • Full mobilization of National Guard and Reserve units to |
| |    — Augment owner/operator security for key telecommunication facilities |
| |    — Support PSN reconstitution |

# Chapter Five

# IDENTIFYING SECURITY TECHNIQUES

In Chapter Three we introduced a set of 20 vulnerability attributes for information systems and their components. Chapter Four presented 13 categories of information system security techniques. We now provide a tool that should help military units and other organizations execute step 4 of the MEII process—identifying security techniques that can mitigate each one of an information system's vulnerabilities. We also discuss factors that should be taken into account in refining the set of security techniques so identified and ways to accomplish step 6 of the MEII process—testing the implemented techniques. The chapter concludes with some observations on the trade-offs that must be made between security and other objectives.

We do not discuss step 5—applying the security techniques—because it cannot easily be treated outside the context of implementation in specific systems. Thus, no generic guidance can be given. We note, however, that units cannot implement all applicable security techniques to reduce their assessed system vulnerabilities. Some techniques address singularities outside a unit's purview or weaknesses in COTS software. In such cases, units will need to communicate their concerns to the appropriate authorities, reduce dependencies on vulnerable elements where possible, and identify workarounds to use in the event of attack.

## MATCHING SECURITY TECHNIQUES TO VULNERABILITIES

The matrix shown in Figure 5.1 arrays the security technique categories against the vulnerability attributes. Each cell is color-coded to

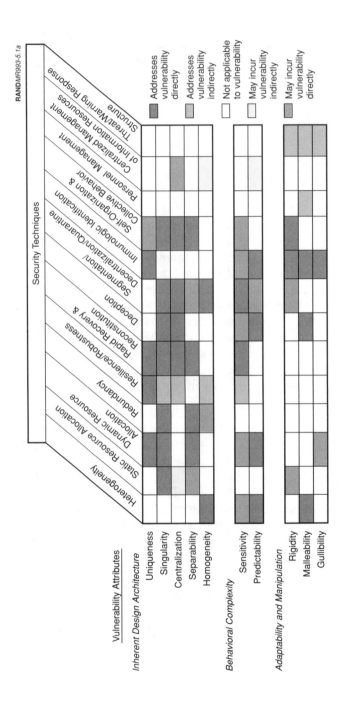

**Figure 5.1—A Matrix Showing the Applicability of Security Techniques to Sources of Vulnerability**

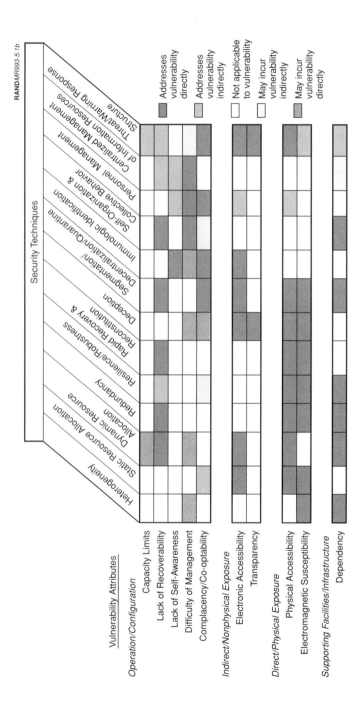

**Figure 5.1—continued**

represent our judgment of the implications of the technique listed above it for the vulnerability given to the left, as follows:

- *Addresses Vulnerability Directly and Substantially (green).* The security technique is applicable to the vulnerability and, if applied properly, can significantly lower it, either by directly addressing the relevant attribute or by greatly reducing the impact of attacks associated with it. These techniques do not necessarily eliminate the vulnerability altogether.

- *Addresses Vulnerability Indirectly or in Modest Degree (light green).* The security technique can help to reduce the vulnerability, but only in an indirect manner, with relatively mild, second-order effects. The technique reduces the impact of an attack to some degree but does not eliminate or address the underlying source of the vulnerability. For such a technique to be effective, it must be used in combination with other techniques that address the vulnerability more directly.

- *Not Applicable to Vulnerability (blank).* It appears that the security technique does not address the vulnerability to any nonnegligible extent.

- *May Incur Vulnerability Indirectly or in Modest Degree (light yellow).* The security technique may increase the vulnerability attribute slightly, or create new problems through second-order effects, such as the psychological responses of system users and administrators. If potential second-order effects are recognized, then additional vulnerabilities of this type may be avoided, or at least kept to a minimal level.

- *May Incur Vulnerability Directly and Substantially (yellow).* The security technique has the potential to create or worsen the vulnerability to a significant extent by directly increasing it or by increasing the impact of attacks exploiting it. The danger of unintended negative consequences can, however, be reduced if the technique is combined with others that address the vulnerabilities that it may incur.

Appendix F provides explanations for the designations assigned to each of the colored cells in the matrix. Because the vulnerabilities and security techniques defined in the two preceding chapters are generic, our judgments of applicability must be notional. The matrix and our evaluations are intended to serve primarily as a framework for thinking about how to improve information system security. The applicability of a particular technique to the vulnerabilities of a specific information system may differ somewhat from our evaluations, depending on the unique circumstances surrounding the architecture, implementation, configuration, and operation of that system.

Let us consider an example of how the matrix works. Two of the principal vulnerabilities of Internet-based information systems are electronic accessibility and malleability. As shown in Figure 5.1, there are seven techniques that address at least one of these two vulnerabilities, but only two that address both: deception and immunologic identification.

Note that each of these techniques has the potential to incur further vulnerabilities. In particular, both of these techniques may incur sensitivity and difficulty-of-management vulnerabilities if they are applied to the system. In addition, deception may incur vulnerabilities associated with complacency and co-optability. This does not mean that deception is necessarily inferior to immunologic identification in this case. It does mean, however, that the nature of the vulnerabilities that would be incurred by each technique should be examined carefully and compared to each other and to the two original vulnerabilities before one or the other is selected.

Let us examine one of the two applicable techniques—immunologic identification—more closely to see how it can help alleviate our two vulnerabilities.[1] As stated in Appendix F, malleability, which may be deliberately designed into a system, has disadvantageous aspects if antagonists gain access to the system as accepted users. In such cases, a pliant operating environment can facilitate an attacker's ability to wreak havoc. If the concepts of immunologic identification could be realized and implemented, this problem could be solved through ubiquitous vigilance. Immunologic identification could

---

[1]Recall that immunologic identification and response in an information system is probably the most speculative and unavailable of the security techniques we describe. Our use of it as an example is for convenience and should not be taken to imply anything more.

recognize irregular behavior and thus determine *when malleability is being abused*, and then trigger an appropriate alarm or countermeasure.

As for electronic accessibility, systems with a great deal of openness absolutely require a broad defensive front: They are likely to suffer a range of attacks that vary in character, space, and time. The plastic, continuous, omnipresent defense potentially offered by immunologic identification could meet that need. Electronic accessibility implies that it is difficult or impossible to prevent a determined attacker from getting in, but through immunologic identification, all interior structures and behaviors could be watched for dangerous foreign or abnormal actions, so that appropriate alarms or countermeasures would be triggered when necessary.

In essence, immunologic identification techniques would address the consequences, as opposed to the source, of vulnerabilities associated with electronic accessibility and malleability. In other words, such techniques would detect and react to attackers; they do not protect the system ahead of attack. Thus, implementing immunologic identification would not directly address these two vulnerabilities, but would greatly reduce the impact of any attacks that try to exploit them.

There are, however, substantial risks associated with employing security techniques of this type. For example, they increase vulnerabilities associated with system sensitivity. Immune agents' sensitivity and autonomy, which are necessary for their effectiveness, may lead to significant overhead and numerous false alarms. Worst of all, they can trigger autoimmune reactions where legitimate users engaging in nonhostile action inadvertently trigger countermeasures. (These countermeasures against legitimate users might also be triggered deliberately by a clever, hostile intruder.) For example, neural net processes are used to recognize anomalous credit card usage. "Anomalous" does not necessarily mean "illegitimate." If a legitimate but anomalous transaction is impeded, it is a mere hindrance; but if the countermeasure is to revoke the card, a good customer will have been cut off.

As suggested by the preceding example, immunologic identification techniques could also increase vulnerabilities associated with difficulty of management. The system would need to be configured correctly, which might be especially difficult initially when the new security features are being integrated into the system. The "immune response" of the system would also need to be monitored and refined constantly, so as to balance security requirements with the efficiency and convenience concerns of legitimate system users. Problems associated with meeting these management challenges would undoubtedly create some unintended flaws or weaknesses, which could conceivably be exploited by a knowledgeable and experienced attacker.

These potential vulnerabilities should be taken into account when considering the benefits of immunologic identification. It might be possible to avoid some or all of them, but that would almost certainly add to the cost and effort associated with implementing such an approach. Ultimately, the vulnerabilities that would be incurred must be compared with the existing ones. Indeed, it may well be that the new problems are just as significant as the old ones, and implementing the security technique is just not worth the time, effort, or cost that it would entail. This type of judgment, of course, depends heavily on the significance of the current vulnerabilities and the details of the specific system under consideration.

## REFINING THE APPLICABLE SET OF SECURITY TECHNIQUES

Suppose that uniqueness has been identified as a vulnerability. As discussed above, vulnerabilities cannot be mitigated by negating them; that is, unique installations cannot simply be eliminated, since they are presumably unique for good reasons (of course, uniqueness should be eliminated where there are *no* good reasons for it). To the extent that this vulnerability is inherent, Figure 5.1 suggests that dynamic resource allocation, resilience and robustness, rapid recovery and reconstitution, and immunologic identification may be applicable techniques.

But before these four candidates can be added to the security technique set for the system under consideration, the security analyst must flesh out how each of them would be applied in this particular case. For example, a combination of administrative procedures and technical solutions might be developed to allow reloading, restoring, and rebooting of unique systems, and special monitoring processes and techniques might be employed to detect unexpected problems arising from system uniqueness. In some cases, the security analyst may decide that a particular technique will not alleviate the vulnerability at hand, even though Figure 5.1 suggests the potential for doing so.

When this process has been completed for all identified vulnerabilities of a system, the result will be a large set of security techniques whose application to this element have been demonstrated by the analyst. The notion of defense in-depth suggests that as many such techniques should be applied at as many different levels and in as many different ways as possible. This should slow an attacker by requiring multiple defenses to be overcome, as well as providing no "safe haven" where an attacker can relax and cease to worry about being detected or defeated. It is also a way of hedging one's bets as to the effectiveness of any single application of any one security technique. Even if that application were to prove not entirely effective, some other technique somewhere else is likely to save the day.

However, the security of no system element should be left to rely on such rescues. As discussed in the context of vulnerabilities in Chapter Three, most systems are built out of—or on top of—other systems, whose vulnerabilities they therefore inherit to some degree. If defense in-depth is truly to be provided, those responsible for system security must recognize this and not assume that lower-level layers or components will necessarily deal successfully with their own vulnerabilities. Any system element that depends on other elements should defend itself against failures of those other elements. This involves minimizing the trust among elements of a system: Each element should protect itself (to the extent possible) from the actions of its subordinate and component elements, even if these elements are normally trustworthy.

Furthermore, security techniques should not simply be applied anywhere and everywhere without regard to their possible interactions. Positive, synergistic interactions among such techniques should be maximized, while negative interactions are minimized. Both synergistic and negative interactions among security techniques may occur in at least three situations:

1. Whenever more than one security technique is identified for a given vulnerability of a given system or element.

2. Whenever a single security technique is chosen to address more than one vulnerability. For example, recall the illustrative analyses in Chapter Three, in which the COE exhibited vulnerabilities associated with uniqueness and that stemmed from the singularity of the kernel for each OS. If techniques of rapid recovery and reconstitution are to address both, they might be implemented so as to mutually reinforce each other.

3. Whenever a given security technique already in use for one system or element is chosen to address a related system or element. For example, might the rapid recovery and reconstitution implemented for a COE OS reinforce the rapid recovery and reconstitution implemented for a COE file system?

This multidimensional analysis should be performed for negative interactions as well as for positive, synergistic ones. Since negative interactions may be quite subtle, it might be helpful to perform an abstract analysis of the full set of security techniques to identify which ones potentially interact with which others in negative ways. Such an analysis would point out where the security analyst should be alert to the possibility of negative interactions, even though this negative potential may not materialize in a given case. This analysis remains to be done but should be relatively straightforward.

Meanwhile, military units and other organizations can apply the process described above. The process should not, of course, be performed mechanically or slavishly: It requires a deep understanding of how various modifications may affect the system at hand and may interact with each other. Nevertheless, we believe that this approach provides a principled way of reducing the vulnerabilities of systems in the information infrastructure.

## TESTING APPLIED SECURITY TECHNIQUES

Constant vigilance will be required to guard against security compromises. But, absent a clear threat, security may take a back seat to other, more pressing day-to-day functions. The only way to keep security in proper perspective is to ensure that the systems are challenged to expose exploitable weaknesses and to fix those weaknesses as they are exposed. There are three approaches to such challenges—threat or attack simulation, "red teams," and invitation of attack by external mischief makers. Such actions represent step 6 of the MEII process set out in Chapter Two.

In military units, the most overt way to challenge systems for exploitable weaknesses is to simulate a threat or the consequences of an attack. U.S. military units must increasingly conduct realistic training exercises that include deliberate "taking down," reduction, or disabling of essential information links and assets, to assure that workarounds and alternatives for these systems have been planned and are effective.[2] To the extent that accidents happen during both training exercises and military operations that disable important information assets, these accidents should be culled for "lessons learned."

More generally, "red teams" can be used to help ensure that information systems will meet security challenges. "Red team" activities can range from threat or attack exercises to critical reviews of security procedures. In the commercial context, red teams will have constraints based on legal and collateral damage considerations. The information system owners and operators will of course have to be an integral part of these red team exercises. Red team activities cannot expose vulnerabilities to potential enemies or be illegal or, for that matter, unduly embarrass information system owners and operators. Acceptable red team procedures and, where necessary, legislation will have to be developed to exploit significant red team potential. The more difficult these legal and procedural constraints become in practice, the more likely security checks will come to depend on "help" from hackers and other external bad actors.

In any event, red teams are unlikely to be effective if they are limited to the occasional, ad-hoc security exercise. There are three ways of institutionalizing red-teaming: (1) creating dedicated red teams, (2) creating a day-to-day "red-team mentality" among normal informa-

---

[2]This approach is now part of DoD's recent Critical Asset Assurance Program, DoD Directive 5160.54 (January 20, 1998). See section 5.4.2: "The Chairman of the Joint Chiefs of Staff shall ensure that disruption and loss of Critical Assets, to include supporting national infrastructures, are scripted and responded to in Joint Exercises."

tion systems operators and users, and (3) getting the unwitting help of hackers and other external threat agents.

On that last point, MEII portions of systems might be left open (in the sense of not hidden) and perhaps even advertised as "invulnerable." Although their advertised invulnerability would probably be an overstatement, it would act as a red flag to hackers worldwide and make those systems a particular target of attack. Through these attacks, vulnerabilities could be gradually located and fixed, to build up resilience and survivability techniques. Properly monitored, such an approach could provide good surveillance and warning regarding the kinds of attack methods that might be expected in the future. Note also that this kind of attack (i.e., testing of system resilience) would be provided for free!

We are aware that the advertising of essential systems as invulnerable is a risky proposition. In fact, it increases transparency (being known to malevolent actors as existing and essential). Our proposal also flies in the face of the protection techniques related to deception. In addition, if vulnerabilities uncovered are hard (or even impossible, in the near term) to fix,[3] there could be great danger in publicizing these vulnerabilities. Perhaps the right compromise is this: *A few* MEII-type systems might be touted as invulnerable, but under carefully controlled circumstances. And these systems should be heavily monitored and isolated from other essential MEII systems. They should be viewed primarily as isolated testbeds and deceptive "honeypots" to attract attackers.

## TRADING SECURITY OFF AGAINST OTHER VALUED ATTRIBUTES

In discussing what security techniques to apply to an information system, we have limited ourselves to implications for system security. Naturally, other factors must be taken into account—cost, of course, and such aspects of functionality as ease of use and efficiency of operation. We cannot here attach values to any one of these system attributes in terms of the others, but we will make a few observations about trade-offs.

---

[3]See Table 3.2 and our discussion about vulnerability distinctions in Chapter Three.

Security per se is not something anyone lobbies against or believes unnecessary. But security comes at a price. There is a dollar price to the security measures identified in Chapter Four—a price that will reflect higher development costs for software, the costs of redundant system elements, the cost of training, and so on. These costs may be sufficient to affect the choice of one security technique over another or the decision about whether to enhance security at all.

We suspect, however, that such decisions will be more strongly influenced by less explicit costs—costs denominated not in dollar terms but in losses of system functionality. Since the purpose of an information system is not to protect itself, security measures cannot seriously compromise a system's ability to fulfill its true purpose or else there is no point in implementing them. Such compromises may take the form of

- slowing the system down so it takes longer to deliver the desired output

- making the system less reliable, i.e., more susceptible to crashes or malfunctions

- making it more difficult for the user to interact efficiently with the system.

Some sacrifices along one or more of these lines may be required to make a system more secure. As pointed out in Chapter Three, the characteristics that make a system easy and efficient to use may also make the system easy and efficient to attack. At some point, however, the marginal gain in security from the next measure applied will be less than the value of the lost functionality incurred. This trade-off between security gained and functionality lost will not be the same for all security techniques, and for any given technique it will vary from system to system. This trade-off thus represents another basis for making judgments as to which techniques to apply in a given situation.

The trade-offs required between security and functionality make a good argument for the implementation of alert levels of the INFOCON type mentioned in the preceding chapter. At lower perceived threat levels, systems might be allowed to function at maximum efficiency in a relaxed security environment. At higher levels, various of our suggested security techniques might be applied more stringently. Examples include instituting dynamic resource allocation, tightening firewalls for greater robustness, instituting special measures to enhance the capability for rapid recovery and reconstitution, greater use of deception, and enhanced immunologic defenses. Note that this is a way of accommodating security-functionality trade-offs, not of avoiding them: maintaining systems at low-alert, reduced-security levels runs the risk that the perceived remote-threat regime does not match reality. (In addition, the necessity of doing something out of the ordinary to invoke more stringent safeguards may work against their being invoked often enough or soon enough.)

More generally, the entire process we propose is not a cheap or easy one. It requires units with varying missions to commit what they may regard as substantial resources to detailed analyses of information function dependencies, system vulnerabilities, and the suitability of various remedies. We suspect that in most cases the gain in security will be worth the cost. As a step toward ensuring that it is, the matrix provided in Figure 5.1 and the method in which it is embedded should ease some of those analyses.

Our efforts, here, however, are only a first step toward elaborating a process for determining essential infrastructures, investigating their potential vulnerabilities, and applying survivability measures. To become a useful, operational plan, our study must be extended and elaborated in at least the following three areas:

- *Security techniques:* Our study concentrated on government-funded research and development in information assurance techniques. There are already, however, many COTS products available today in the areas we examined. These include intrusion detection systems, firewalls, and encryption schemes. Commanders and organization leaders need advice on the type, cost, and quality of products that can be procured and applied today in each of our security categories; they cannot wait for today's R&D to become available products sometime in the coming decade.

- *Cost-benefit analysis:* For a particular information infrastructure, a vulnerability in each of our 20 listed categories carries a certain level of risk to an organization. For that organization's information system, applying various techniques within our 13 security categories entails certain costs. Commanders and organization leaders should be provided the means to work through a type of cost-benefit analysis—even at a qualitative level—to aid in deciding which vulnerabilities are most important and which security techniques can be afforded to ameliorate them.

- *Validation of the methodology in the real world:* Our analysis to date has been quite abstract, even theoretical. It should be applied, as a case study, to the analysis of a real-world essential information infrastructure, to see if its use results in a deeper understanding of vulnerabilities and applicable security techniques than would otherwise have been achieved.

# Chapter Six

# DISTRIBUTION OF RESEARCH EFFORT

If the potential of the security techniques identified in Chapter Four is to be realized, further research will be necessary—if for no other reason than to keep pace with the vulnerabilities of the evolving information infrastructure. A comprehensive, accurate assessment of research needs would require a comparison of the state of the art in each of our security technique categories with that category's potential and the likelihood that future research would help make up the difference. That is a research project in itself. We believe, however, that we can take a first step and draw some suggestive inferences by simply determining the evenness with which a sample of recent research projects are distributed across the security technique categories.[1]

## APPROACH

We examined relevant research funded by two key U.S. government agencies: The Defense Advanced Research Projects Agency (DARPA) and the National Security Agency (NSA). These agencies were chosen because of their tradition of funding "leading edge" computer security research; their agendas are thus a good indicator of trends in research focus. This is not a representative sample but it does constitute a substantial fraction of the publicly funded research in this area.

---

[1]For our sampling of research projects, we limited ourselves to the 11 technologically oriented security technique categories shown in Table 6.1 and Figure 6.1, plus a category for other/miscellaneous.

Specifically, we reviewed 104 projects funded in fiscal year 1998 by the Information Survivability Program within DARPA's Information Technology Office (ITO) and another 6 funded in fiscal year 1998 by the Information Assurance program within the same agency's Information Systems Office (ISO).[2] We included 45 NSA information security projects.

We also examined a nonstatistical sampling of 20 research projects at universities, national laboratories, and in the private sector that are using "biomimetic" approaches. Among the research programs reviewed were those of Forrest et al. (1994) at the Santa Fe Institute and Kephart et al. (1997) at IBM's AntiVirus group. Their studies address the range of biological approaches, including network epidemiology, genetic programming, autonomous agents, and computer immune systems.

Some explanation of the biomimetic approach may be helpful here (for more detail, see Appendix C). By "biomimetic research," we mean avenues of study whose goal is to explicitly replicate biological principles in other fields. The abstraction of principles from living systems has already demonstrated its value, for example, in artificial intelligence efforts using neural networks. With respect to an MEII, we note that organisms exhibit flexibility in meeting uncertainty and novel hazards—a tremendous boon in a rapidly changing environment. This flexibility has evolved over long periods of time and through a variety of ecological contexts. We may thus have something to learn from biological characteristics in thinking about improving the security and survivability of information systems, particularly considering similarities between biological and information systems in complexity, interdependence, and adaptiveness. In the application of biological principles to information system security and survivability, there have been three main avenues of exploration to date:

- *Immune System Defense Models.* These projects seek to capitalize on mammalian immunology, which is perceived as flexible,

---

[2]Summary information for most of the DARPA projects is available at http://www.darpa.mil/ito/research/is and at http://web-ext2.darpa.mil/iso/ia/.

continuous, distributed, autonomous, wary, adaptive, and very successful against a range of both known and novel adversaries.

- *Insect Models of Organization.* Ethological research on the self-organizing, collective resilience of ant and other insect societies has led to research efforts seeking to emulate their successes.

- *Diversity-Based Approaches.* Heterogeneity, from both anatomic and functional standpoints, is seen as essential to a species' survival. Research along these lines includes analyzing the epidemiology of damaging vectors across networks and the manner in which complication and diversity can impede attackers, along with optimizing software code through "genetic programming."

Projects within these categories are at widely varying states of application: In the case of immune system models, commercial products are already available, while insect models remain in the research and development stage. We include a selection of the work being done, chosen for its representativeness and salience.

## RESULTS

Based on the titles and the summary information (which was fairly limited and often quite technical in nature) on the various projects described above, we took a first cut at placing each of them into one or more of the security technique categories described in Table 6.1.[3] For more specificity, we assigned the projects to subcategories based on the research theme or approach. Several projects were assigned to multiple categories. In the table, projects are coded "T," "S," "N," and "B," according to whether they are funded by DARPA ITO, funded by DARPA ISO, funded by NSA, or take a biomimetic approach. The numerical portions of the project codes are keyed to a list in Appendix G that provides project titles, performing institutions, and principal investigators. Table 6.1 shows the total number of projects in each category and subcategory. The category totals are graphed in Figure 6.1.

---

[3]Personnel from NSA assisted us in categorizing that agency's projects into the various security technique categories.

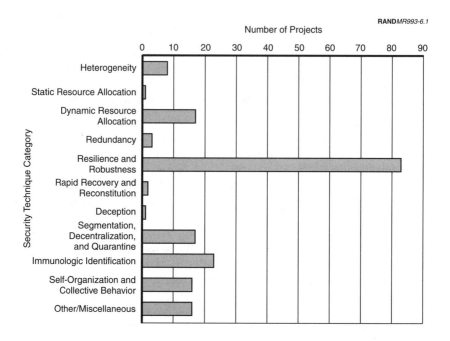

Figure 6.1—A Distribution of All Research Projects Examined

Two important observations can be drawn immediately from this review.  First, the largest number of projects, by far, is in the resilience-and-robustness category.  This may initially be neither surprising nor worrisome, since resilience and robustness comprise a widely applicable set of security techniques.  However, as discussed in Appendix F, increasing resilience or robustness tends to have only a secondary effect on most types of systems; both reduce the consequences of attacks rather than the vulnerabilities themselves.

Second, four categories are addressed by no more than three projects each.  Those categories are static resource allocation, redundancy, rapid recovery and reconstitution, and deception.  Most static resource allocation and redundancy technologies are fairly simple and unsophisticated, so it is not particularly worrisome that DARPA, NSA, and the biomimetic researchers are not focusing on these types of techniques.  On the other hand, rapid recovery and reconstitution together with deception can provide potentially useful means of

Table 6.1

Categorization of Research Projects

| Security Technique Subcategory | Total | Codes of Projects Assigned |
|---|---|---|
| *Heterogeneity* | 8 | |
| Preferential replication/lifespan | 3 | B3, B4, B19 |
| Architectural/software diversity | 2 | T49, B17 |
| Path diversity | 1 | B16 |
| Randomized compilation | 1 | B15 |
| Secure heterogeneous environments | 1 | T50 |
| *Static resource allocation* | 1 | |
| Hardware technology | 1 | N4 |
| *Dynamic resource allocation* | 17 | |
| Detect and respond to attacks/ malfunctions | 11 | T4, T5, T18, T21, T32, T91, T104, S5, S6, N20, N21 |
| Dynamic quality of services | 2 | T15, T84 |
| Active packet/node networks | 2 | T30, T65 |
| Dynamic security management | 2 | *T2*, N19 |
| *Redundancy* | 3 | |
| Replication | 3 | T52, T89, T103 |
| *Resilience and robustness* | 83 | |
| Cryptography/authentication | 30 | T7, T11, T23, T25, T31, T33, T36, T37, T44, T67, T68, T71, T80, T83, T96, T97, N1, N3, N5, *N14, N15, N16, N17, N18, N33, N34, N35, N38, N39,* N45 |
| Modeling and testing | 12 | T1, T8, T19, T20, T47, T48, T69, T77, T78, N10, N11, N42 |
| Fault/failure-tolerant components | 11 | T3, T38, T53, T58, T60, T62, *T66,* T88, N2, N6, N7 |
| Advanced languages and systems | 9 | T13, T14, *T26a, T26b,* T29, T51, T54, *T101,* N31 |
| Wrappers | 6 | T22, T28, T55, T74, T100, T103 |
| Firewalls | 5 | T70, T72, T76, T79, N12 |
| Secure protocols | 7 | T73, T90, S1, N30, N40, N41, N43 |
| Advanced/secure hardware | 3 | T59, *N36, N37* |
| *Rapid recovery and reconstitution* | 2 | |
| Detect and recover activities | 2 | T6, T57 |
| *Deception* | 1 | |
| Decoy infection routines | 1 | B18 |
| *Segmentation, decentralization, and quarantine* | 17 | |
| Secure distributed/mobile computing | 6 | T39, T56, T63, T86, T87, N32 |
| Enclave/shell protection | 5 | T76, T79, T102, S1, N13 |
| Intruder detection and isolation | 2 | T9, T10 |
| Specialized "organs" | 2 | B6, B7 |
| Autonomous self-contained units | 1 | T75 |

## Table 6.1—continued

| Security Technique Subcategory | Total | Codes of Projects Assigned |
|---|---|---|
| Damage containment | 1 | <u>B8</u> |
| *Immunologic identification* | 23 | |
| Autonomous agents | 4 | <u>T23</u>, T45, T64, T92 |
| "Lymphocyte" agents | 6 | B2, <u>B3</u>, <u>B4</u>, B5, <u>B6</u>, <u>B7</u> |
| Detection of anomalous events | 4 | <u>T5</u>, T16, T17, B20 |
| Mobile code verification | 4 | T27, T35, T41, N9 |
| Self/nonself discrimination | 3 | T40, T42, B1 |
| Information dissemination | 2 | <u>B8</u>, B9 |
| *Self-organization and collective behavior* | 16 | |
| Adaptive mechanisms | 5 | T34, T43, T81, T85, T93 |
| Formal structure modeling | 3 | T12, T24, T61 |
| Emergent properties and behaviors | 3 | B10, B12, B13 |
| Node/software optimization | 3 | B11, B14, <u>B19</u> |
| Market-based architecture | 1 | T94 |
| Scalable networks (VLSI) | 1 | T99 |
| *Other/miscellaneous* | 16 | |
| Multiple approaches to network | 12 | T46, *T82*, T95, *S2*, *S3*, *S4*, N8, N22, N23, *N25*, *N27*, N44 |
| Security/survivability technology forecasting | 4 | N24, *N26*, *N28*, *N29* |

NOTES: The letter in each project number indicates the source from which we drew the project: T = DARPA ITO; S = DARPA ISO; N = NSA; and B = biomimetic research. Projects that have been assigned to two categories are underlined, and those for which summary information was not available are in italics.

assuring the survivability of an information system. This is especially true of deception, which probably deserves considerably greater attention in future research efforts—if the dearth of deception projects in our sample is representative of the broader research environment. Appendix E provides a more thorough examination of how deception could be employed to effectively protect information systems from attack.

# Chapter Seven

# RECOMMENDATIONS

A number of recommendations are implicit in the conclusions in the previous chapter. We emphasize here some overall actions we feel can be taken on the basis of our findings to date.

## USE OUR METHODOLOGY AS A CHECKLIST

As mentioned previously, we believe that DoD operating units, government agencies, and commercial industry groups should *not* expect there to be a hardened, secure backbone minimum essential information infrastructure on which they can depend. The needs are simply too geographically and functionally diverse. Rather, each unit commander or leader, at varying levels in organizational hierarchies, should consider using our methodology as a type of checklist: Which functions are essential in his or her unit's operation? Which information systems are essential for that functionality? Which of our 20 types of vulnerabilities do those systems exhibit, at various architectural levels? Which of our 13 categories of security techniques are relevant in making those systems more survivable? How can any additional protections that are implemented be tested against various attack scenarios?

## DEVELOP A TEST SET OF SCENARIOS INVOLVING IW ATTACKS

In developing various protection strategies for greater system survivability, a set of attack scenarios is needed against which those strategies can be tested. Currently, a great deal of emphasis is being placed on unstructured, uncoordinated hacker threats to information systems and information infrastructures. Study should be made of what theoretically could be accomplished by a well-funded series of coordinated probes, intrusions, or attacks on information systems and information infrastructures. The focus should be on how postulated nation-states, transnational terrorist groups, and nonstate actors might perform cyberspace reconnaissance and attacks on information systems and information infrastructures to accomplish their objectives. What could a patient and determined intruder accomplish over a long period of time without being easily detected? The intruder activities might include the covert use of hacker tools as well as a wide range of insider techniques to accomplish specific long-term objectives. With such scenarios available, many of the DoD, government, and commercial unit leaders could then test their essential systems' robustness and survivability against worst case attacks.

## DEVELOP CASE STUDIES OF OUR PROPOSED METHODOLOGY

In Chapter Three we presented some brief case studies involving the existing architectures of systems such as satcom communication and the Windows NT operating system. Much more thorough, detailed investigations of the design, implementation, and use of essential U.S. information systems should be conducted to see

- if new categories of vulnerabilities are present that we have not uncovered to date

- if new security techniques suggest themselves

- whether there are critical vulnerabilities for which security seems unavailable in systems of national strategic importance

- how useful our proposed methodology is as a framework for conducting such analyses.

## EXPLORE BIOLOGICAL ANALOGIES IN MORE DETAIL

In our work, we have been impressed with the usefulness of biological analogies at various levels (see Appendix C) in suggesting security techniques that aid complex systems in achieving survivability under a wide variety of circumstances.  More intensive study of these analogies and their implications for system design should be undertaken by research teams combining skills in genetic, biological, computer, and information system specialties.

## CONSIDER R&D ON SURVIVABILITY TECHNIQUES TO "FILL THE GAPS"

Our analysis in Chapter Six indicated that much of the current government-funded R&D on information survivability is clustered in some of our categories of security techniques, but other categories, such as deception and rapid recovery and reconstitution, seem to be relatively ignored.  Those underfunded security categories might be examined to see if useful tools, techniques, and strategies might be developed in those areas to aid in their application in a broad national campaign designed to increase essential information system survivability.  We believe the general topic of deception, in particular, is a rich lode to be mined.

# HISTORICAL NOTE ON THE U.S. MINIMUM ESSENTIAL EMERGENCY COMMUNICATIONS NETWORK (MEECN)

The cold war predecessor of the MEII concept was the Minimum Essential Emergency Communications Network (MEECN). The purpose of the MEECN was to assure the timely receipt of emergency action messages (EAMs) that initiate execution of the Single Integrated Operational Plan by worldwide U.S. nuclear forces under nuclear attack by the Soviet Union. Day-to-day communications between the command authorities and U.S. nuclear forces were dependent on military and commercial systems that were either not expected to survive nuclear attack or that had unacceptable performance uncertainties in nuclear-effects environments. For example, it was known that high-frequency communications are subject to blackout and land-line (PSN) communications depend on a relatively small number of fixed, targetable switching stations and net control nodes.

The dependence of strategic communications on the PSN was particularly troublesome. The problem was that the damage potential from various possible modes of attack could never be reliably assessed. Further, as a consequence of these uncertainties and because of the potential expense of hardening the PSN systems "just in case," alternatives had to be developed and fielded. The alternatives to this and other strategic communications problems evolved over time into the MEECN.

The MEECN was a dedicated overlay on day-to-day communications systems such as the PSN. It consisted of command, control, and communications relay aircraft (PACCS, TACAMO), missiles with ultra-high-frequency (UHF) broadcast systems (ERCS), specialized satellites (MILSTAR), and ground radio broadcast systems (e.g., GWEN). It was not a hardened subnet of the PSN. It was formed from rather different systems (e.g., very-low-frequency transmitter and airborne UHF relay) that were independent of the uncertainties in the PSN. It was a special-purpose, dedicated, hardened, high-confidence emergency system. The MEECN mission was to pass the EAMs (short, formatted alpha-numeric codes).

# HOW THREATS RELEVANT TO AN MEII DIFFER FROM HACKER NUISANCE ATTACKS

The potential threats we are most concerned with in this study have certain properties that distinguish them from the more common hacker nuisance attacks. First, the attacker ought to be able to *predict the consequences* of the information system attack fairly reliably. Second, these consequences should *measurably increase the likelihood that the attacker's strategic objectives will be met*. Finally, information system attack options must *compare favorably with alternatives* that might achieve the same sort of results.

Predictability will be important if the attacker's objective is not simply to cause trouble but to embed information system attacks effectively in a larger campaign involving other elements, including more conventional political, economic, and even military operations. If the information system attacks are essential to the success of other strategic elements in the attacker's plan, they had better work as intended if the strategic objectives are ambitious (i.e., if the consequences of failure could be dire). What could go wrong? At one extreme, the information attacks could be detected and the perpetrators identified, providing warning that could be used to upset other elements of the attacker's plan. At the other extreme, the information attacks could be "too successful," resulting in collateral damage that escalates the conflict beyond the attacker's intent, potentially causing him or her great harm. Concerns of the first sort typically limit the military applications of special-operations forces and other covert operations. At the other extreme, if serious damage were done to, for example, power distribution in the United States, causing significant loss of "innocent" lives, the U.S. response might be extreme.

The second critical threat feature is that the effects of the attack should bear some important relationship to the rest of the overall plan. This is where threat discussions become complicated (and contentious) because they require much scenario speculation. Given a scenario, the effects of the IW components still may be difficult to assess. This is even true of the relatively straightforward assessment of the value of $C^4ISR$ degradations in electronic warfare scenarios (e.g., how does the jamming or physical destruction of a specific early-warning radar really affect the course of the war?).

Finally, there may be several means to reach the same ends. Assume, for example, that the attacker's goal is to slow the deployment of U.S. troops to his or her region by several days, thereby presenting the United States with a much more difficult problem to solve (e.g., forced entry following an invasion). He or she could then disrupt the computers and communications used to manage the deployment stateside or infiltrate a few small teams with shoulder-fired infrared homing surface-to-air missiles to destroy a few transport aircraft just after takeoff. That might cause a complete halt to air traffic until the teams could be found and neutralized. Which approach might be better is open to discussion.

# BIOLOGICAL ANALOGIES FOR INFORMATION SYSTEM SURVIVABILITY

Evolutionary processes have created biological systems that are survivable in an uncertain world. This survivability is exhibited at many levels of resolution: Nucleic acid, cell, organ and system, individual organism, and population. Information systems evolve, aggregate, replicate, interact, and adapt in ways often eerily reminiscent of organic entities. This is sometimes a deliberate design effect. We, and others, are therefore interested in what can be learned from biological examples and metaphors that can aid in obtaining greater survivability of information systems in a hostile and fluid environment. To this end, we have studied research now under way on biological principles for information system security and have searched for additional analogies between the biological world and that of complex information systems. We have concluded that these metaphors are useful in extending our thinking into new areas. The categories of bio-inspired survivability techniques we have developed aided us in composing the list of "security techniques" discussed in Chapters Four and Five. We believe further study of biological analogies might be helpful to the field of information system security, survivability, and defensive information warfare (IW-D).

## CONTEXT AND PURPOSE

The very moment we speak of survivability, we speak in biological terms. And biology is nothing if not infinitely illustrative in the depth and breadth of its inventiveness for assuring survival. Given the scope of variability in the environment of an organism or population, protective, detection, and reactive measures must be flexible and effective under myriad conditions, inclement and idyllic alike. Moreover, such survivability techniques must hold up even when "typical" change is occasionally punctuated by wholly unforeseen environmental disasters and accidents.

If only weather were the worst thing to worry about. Predators and competitors are a constant and mortal threat to be avoided, fended off, endured, or destroyed. Organisms additionally have to worry about prey, too: The availability and abundance of one's own dinner is as essential to survival as not becoming someone else's entrée. Here too, examples abound of inventive methods for preserving the integrity and functionality of organisms, from their smallest architectural components to their largest social groups.

It is with the above considerations in mind that we began exploring the value to be added in examining selected methods by which biological infrastructures remain viable in a dynamic, often inimical, environment.

As introduced in Chapter Four, we have defined categories of information infrastructure security techniques that were drawn from "classical" information systems, fault-tolerance, and computer security concepts. Further, as noted in Chapter Four, it was the cycles of feedback from abstracting these categories to biological infrastructures that in part helped us to refine our definitions. In undertaking the analysis reported in this appendix, we thus had three objectives:

- Evaluating the efficacy of traditional information infrastructure protection and survivability concepts *against a biological backdrop* as a way of gauging their generic applicability and robustness.

- Refining and amending our categories in areas where biology proved enlightening and compelling.

- *Mapping back to information technologies from biology* those defensive measures seen to be widespread and effective in nature, yet missing in information infrastructure protection.

To achieve the first of these objectives, we needed to look for examples where such protective/survivability measures were mirrored in biology, specifically as relates to preserving infrastructure. We did so at each of the five levels of resolution or aggregation mentioned above, from nucleic acids through populations of organisms. We sought to answer, for the first ten protection technique categories,[1] the question, "Is there at least one biological example of this at each level of resolution?" If we found no biological model for our technique, we reexamined it conceptually, perhaps refining our description of the attribute, or subsuming it into another category.

It turned out to be virtually impossible for us to find examples of information infrastructure protection that had no analog in biology. Yet we were able to hone and clarify our definitions through this exercise. For example, stemming from the scrutiny of biological cases, we drew a distinction between armoring a system to preclude damage (static resource allocation), and toughening a system to withstand damage (resilience and robustness). It is our opinion that this distinction should be recognized by designers and administrators of information systems.

Lastly, in the course of examining the survivability/protection issues, we found biological principles that seemed relevant and apt, yet with no clear parallel in "classical" information infrastructure protective measures. In those cases, we mapped backwards to the domain of information infrastructure defense. The single best example of this is the category of immunologic identification, with its notions of self-awareness, memory, learning, flexible detection, adaptation, continuous and ubiquitous function, etc.

## ANALOGIES

We have telegraphed the outcome of our analysis. We found that the categories introduced in Chapter Four are widely and readily observable in nature as methods of protecting and preserving biological infrastructures. The particulars varied frequently, as will be visible below, but the guiding principles remained clear. For each of the following protection, detection, or reaction techniques, we volunteer

---

[1]We omit here the three categories based largely on human organizational considerations: Personnel management, centralized management of information resources, and threat/warning response structure.

what we consider to be an illustrative example of its application at each biological "level of resolution."

## Heterogeneity

Heterogeneity may be spatial or temporal:

*Examples:*

- Nucleic acids:  the multiplicity of DNA-repair enzymes (glycosylases, "bulky lesion" repair enzymes, etc.)

- Cells:  the diverse types of naturally occurring analgesics—endorphins, enkephalins, etc.

- Organs and systems:  the numerous and differing defensive measures in the gastrointestinal (GI) tract (low pH, vigilant epithelia, etc.)

- Individual organisms:  sexual reproduction as a robust means for ensuring future diversity

- Populations of organisms:  SOS response in *Escherichia coli*—secondary effects of increasing mutation rate.

## Static Resource Allocation

The a priori assignment of resources preferentially, as a result of past experience or perceived threats, with the goal of precluding damage.

*Examples:*

- Nucleic acids:  maintaining "junk" DNA to absorb damage

- Cells:  sequestering and maintaining chromosomes in the nucleus

- Organs and systems:  organ placement beneath bone or muscular sheath

- Individual organisms:  protection of entire animal by exoskeleton (e.g., arthropods) or appropriation of extraneous protection (e.g., hermit crabs)

- Populations of organisms:  fortification or stockpiling.

## Dynamic Resource Allocation

Some assets or activities are accorded greater importance as a threat develops.  This technique calls for directed adaptation to inclement conditions as they evolve.

*Examples:*

- Nucleic acids:  SOS response in *E. coli*—DNA repair primacy
- Cells:  heat shock response (cascade effects)
- Organs and systems:  blood shunt (cardiopulmonary precedes GI tract in import)
- Individual organisms:  sympathetic nervous response ("fight or flight")
- Populations of organisms:  hive repair.

## Redundancy

Maintaining a depth of spare components or duplicated information to replace damaged or compromised assets.

*Examples:*

- Nucleic acids:  diploidy
- Cells:  multiple cell cycle checkpoints (preventing transformation)
- Organs and systems:  pluripotency of stem cells for replenishment and regeneration
- Individual organisms:  bilateral symmetry
- Populations of organisms:  division of labor with auxiliaries.

## Resilience and Robustness

Ability to rebound to the status quo ante after a perturbation. Sheer toughness—remaining serviceable while under attack, while defending, or when damaged.

*Examples:*

- Nucleic acids: DNA instead of RNA as information "repository"
- Cells: lipid bilayer as cell membrane in eukaryotes
- Organs and systems: lungs (functionality even at seriously degraded state)
- Individual organisms: distributed vitals (e.g., starfish) to avoid mortal wounds
- Populations of organisms: bacterial spore formation (e.g., anthrax).

## Rapid Recovery and Reconstitution

Quickly assessing and repairing damaged or degraded components, communications and transportation routes.

*Examples:*

- Nucleic acids: DNA polymerases (using information from uncompromised copy)
- Cells: heat shock cascade (damaged components rapidly removed, new ones synthesized)
- Organs and systems: injury response (vasospasm, clotting, increased heart and respiratory rate)
- Individual organisms: load-shifting (e.g., neural-plasticity)
- Populations of organisms: activation of certain conifer seeds by fire.

## Deception

Artifice aimed at inducing enemy behaviors that may be exploited.

*Examples:*

- Nucleic acids:  viral DNA insertion (disinformation and co-option)
- Cells: enveloped viruses (e.g., influenza)—a disguise
- Organs and systems: tonsils—a putative honeypot (lure)[2]
- Individual organisms: camouflage
- Populations of organisms: asymmetric strategies.

## Segmentation, Decentralization, and Quarantine

Distributing assets to facilitate independent defense and repair. Containing damage locally and preventing propagation of damaging vector.

*Examples:*

- Nucleic acids:  nearly every cell contains entire genome in nucleus
- Cells: cytotoxic T-cells induce apoptosis of compromised cells
- Organs and systems: distributed immune assets for independent action
- Individual organisms: blood-brain barrier
- Populations of organisms: "sacrifice" of older or weaker individuals to predators.

## Immunologic Identification

Four specific characteristics of the immune system are

- self/nonself discrimination
- partial matching algorithms (flexible detection)

---

[2]Note that this is an unproved hypothesis.

- memory and learning
- continuous and ubiquitous function.

*Examples:*

- Nucleic acids:  antibody generation
- Cells:  cell surface glycoproteins (MHC) as "self" passcode
- Organs and systems:  acquired immunity (B-cell mediated secondary response)
- Individual organisms:  immunologic memory
- Populations of organisms:  recognition of other individuals as members of same or different species.

## Self-Organizing and Collective Behaviors

Valuable defensive properties can emerge from a collection of autonomous agents interacting in a distributed fashion.

*Examples:*

- Nucleic acids:  protein folding for optimal enzymatic action
- Cells:  neural pathway development
- Organs and systems:  concomitant vascularization during organogenesis
- Individual organisms:  cellular slime molds, various colonial organisms (e.g., Portuguese man-of-war)
- Populations of organisms:  insect hive organization.

# PRIORITIZATION IN INFORMATION SYSTEMS

In both military and commercial information systems and networks, it is often desirable or even essential to provide different grades of service for different classes of "customers." This is often called *prioritization*, or, in military systems, *precedence.* (We use the word *customer* here in both its conventional sense and in the broad queueing-theoretic sense.) In assessing the minimum essential information infrastructure for any organization or unit, it is vital to understand the options for prioritization in use of information assets, since various of those assets may be restricted or unavailable in an adverse contingency. Some guidelines in considering prioritization options are given below.

Before proceeding further, we briefly introduce some necessary terminology:

- A *queueing system model* describes a system in which customers wait for some type of service, receive service, and then leave the system. A *queueing network model* is an interconnection of two or more queues such that customers who leave one queue may enter another queue or leave the system.

- A *customer* is anyone or anything requiring access to resources that are not dedicated to it exclusively. In models of circuit-switched communications networks (such as the nonsignaling part of the public switched network), a customer is typically a request for a circuit. In packet-switched communications networks such as the Internet, a customer might be a message or one of its constituent packets. In a model of a multitasking computer system, a customer might be a process, or an interrupt generated by a process. In a queueing model of a barber shop, a customer is a human being.

- A *service discipline* is a rule that determines the order of service for customers waiting in the queue. The simplest and most common service discipline is first-come, first-served, also known as first-in, first-out, but many other service disciplines are possible.

- Some queues have a limit on the number of customers who can wait. In this case, an arriving customer who finds the system full departs without service; this is called *blocking*. Some queues do not permit any waiting, so that arriving customers either enter service immediately or are blocked; such a queue is called a *loss queue*.

## WAYS TO PROVIDE PREFERENTIAL SERVICE

On a conceptual level, there is a fairly small set of basic ways to provide preferential service in a queueing system or queueing network:

1. In queueing model systems with waiting, higher-priority customers can advance in the queue so that they jump in front of any lower-priority customers (customers having the same priority level would presumably still be served in a first-come, first-served manner). A customer who has already entered service is not affected by the arrival of a higher-priority customer. This is known as *non-preemptive priority queueing* or *head-of-line queueing*. In store-and-forward networks, it is desirable to have shorter waiting times for

   — control packets (packets that carry information about the status of the network)

   — packets associated with two-way (interactive) voice connections

   — packets associated with urgent messages (especially in military networks).

2. The service of a lower-priority customer can be interrupted (and later resumed) or terminated to expedite the service of a higher-

priority customer. These are called *preemptive resume-priority queueing*, and *preemptive abort-priority queueing*, respectively. In preemptive resume-priority queueing, a customer whose service is interrupted would typically retain his or her position in the queue, resuming service after the higher-priority customer completes his or her service. Preemptive priority queueing is an important technique in real-time computing systems.[1]

3. Some resources can be set aside or reserved for the exclusive use of one or more classes of users. This technique is sometimes called *resource fencing*. Suppose, for example, that two classes of users share a satellite system with 20 voice channels, and that the high-precedence users account for 20 percent of the total traffic. If the blocking probabilities are judged to be too high for the higher-priority users, 10 channels for each user class might be reserved to ensure that the high-precedence users will have a low probability of blocking. Note that the high-precedence users would thus own 50 percent of the channels, although they generate only 20 percent of the traffic. There are many possible variations on resource fencing, e.g., fewer channels might be allocated to the high-precedence users, but they may be permitted to also access the other channels (if available) whenever all of their own channels are in use. From the standpoint of efficiency, this approach is generally attractive only when high-priority users must receive prompt service but low-priority users cannot be preempted. Whether such bandwidth fencing is advisable from a survivability standpoint will be addressed later.

4. In a system with multiple types of services or resources, one might impose *service type restrictions*, i.e., restrictions on the types of services and resources that can be accessed by low-priority users.

*Example 1:* In a satellite network for mobile users, the lowest-priority users might be permitted to send data and text messages only, the medium-priority users to also send store-and-forward

---

[1]When practical, interruption is generally preferable to termination, both from the viewpoint of the lower-priority user whose service is affected, and from the standpoint of overall system efficiency. However, in the example of the real-time computing system, resuming execution of a process from the point where it was interrupted requires its state to be saved, but this might take more time than can be afforded.

voice, and the highest-priority users to send two-way interactive voice and video.

*Example 2:* In a circuit-switched network, limits might be imposed on connection holding time, with a longer maximum duration for the higher-priority users (or perhaps no limit at all).

Methods 1–3, above, demonstrate the following rather intuitive truth: In both the packet-switched and circuit-switched systems, shorter waiting times and lower blocking probabilities can be provided for higher-priority customers, but this comes at the price of longer delays or increased blocking probabilities for lower-priority customers. Less obvious is the fact that some of these techniques can potentially reduce the overall performance of the system. Resource fencing, for example, can increase blocking probabilities for all classes of users. Thus, it is not necessarily true that a loss of performance for one class of users improves the performance for the remaining users.

The four basic techniques can be combined in various ways. Furthermore, the designers of a network might choose to implement some of the above measures on a dynamic, as-needed basis. For example, designers might choose to implement service restrictions only when the system is experiencing very heavy loads (congestion) or some other stress (e.g., an information attack).

There is an additional type of prioritization that overlaps all of the above types, but it is also distinctive. Because this additional type of prioritization is particularly important for this study, we choose to call it out separately:

5. Consider a system involving multiple shared resources that are subject to accidental or intentional disruption. Almost any such system can be designed so that some "customers" will experience more reliable service (access to whatever resources remain functional). This type of prioritization, which we will call *unequal protection*, can take many forms:

   *Example 1:* Many satellite service providers offer transponder leasing arrangements that guarantee the availability of a backup transponder unless the entire satellite fails (such leases are of course more expensive than leases without a backup).

*Example 2:* Under the Telecommunications Service Priority Program, North American interexchange service providers try to achieve faster service restoration times for some classes of users (e.g., military bases) in the event of a major outage.[2]

*Example 3:* Some telecommunications service providers offer diverse route leases, i.e., a leased point-to-point connection that makes use of two (supposedly) disjoint routes to minimize the chance that a failure in the system disrupts the connection.

*Example 4:* When switching or trunking capacity is lost because of a failure or congestion, one possible response is traffic shedding. In the PSN, this is typically done in a rather indiscriminate fashion, but in a system where traffic is divided into multiple priority classes, traffic could initially be shed from the lower-priority classes.

Although prioritization at the user level is more prevalent in military communications networks than in commercial networks, most commercial networks already make use of prioritization for lower-level functions such as network control, status monitoring, handover coordination, and the like. As noted before, control packets in the Internet receive expedited service. In the PSN, signaling information is passed via a separate, dedicated network; this could be considered a form of resource fencing. Prioritization at the user level is already present in some commercial communications systems and networks, and will become more prevalent in the future as service providers seek to differentiate themselves by offering special classes of services for customers (businesses, the military) who desire better performance or higher reliability and are willing to pay somewhat more for enhanced services.

## PRIORITIZATION AND SURVIVABILITY

Prioritization is an important tool for enhancing system survivability. Although unequal protection is the form of prioritization most obviously suited to the MEII concept, all of the five types of prioritization have some applicability.

---

[2]No data are available on the effectiveness of this program.

If implemented without care, certain forms of prioritization can reduce overall system efficiency or performance, and even possibly reduce the survivability of a system. An example of how this can happen, is the technique of resource fencing as applied to military communications satellites. Suppose that a "bent pipe" (nonprocessed) satellite transponder is divided into 40 equal-bandwidth channels, and that these are assigned to users on a demand-assignment, multiple-access basis. A small mobile jammer might not have sufficient power to be effective against the entire transponder (or might be limited in bandwidth so as not to be able to cover the entire frequency band of the transponder). Suppose that the jammer cannot jam more than 5 of the 40 channels. One possible strategy for an enemy equipped with such a jammer is to randomly select 5 channels, perhaps making a new random selection every 10 to 30 seconds. If, however, the enemy knows that 5 of the 40 channels are reserved for critical traffic, then a more effective strategy would be to sit on these 5 channels.

## AN UNDESIRABLE CONSEQUENCE OF HEAD-OF-LINE PRIORITIZATION

For purposes of illustration, let's consider a simple M/G/1 queueing system[3] with nonpreemptive priorities. Suppose that there are $n$ priority classes, with customers in class 1 having the highest priority and those in class $n$ having the lowest priority. Assume that customers of class $i$ arrive as an independent Poisson process with rate

---

[3]Queueing models are widely used as the starting point for mathematical analyses and simulations of communications networks, including both circuit-switched networks and store-and-forward networks. The entities that move through queueing models are called "customers." These customers are packets, messages, or cells (an ATM packet is called a cell) in a store-and-forward communications network model, or requests for connections in a circuit-switched network model. The simplest queueing model that has any predictive value for real-world systems is the "M/G/1" queue. The "M" in this notation stands for "memoryless" and indicates that the distribution of interarrival times is exponential (equivalently, the arrival process is Poisson). The "G" stands for "general" and indicates that the distribution of service times—i.e., the duration from the moment that a customer reaches the head of the queue and begins service until the completion of that service—is arbitrary (i.e., it is not restricted to be exponential). The "1" indicates that there is only a single server. The M/G/1 queue could be applied, for example, to a statistical multiplexer in which packets from several sources arrive according to Poisson statistics, are stored in a common queue (buffer), and transmitted over a common link to some destination.

$\lambda_i$ and that the mean and mean-square service times are $\overline{X}_i$ and $\overline{X}_i^2$, respectively. Assume that there is no queue limit ($m = \infty$). Under these assumptions, the mean waiting time for customers of class $k$, $W_k$, is given by the following formula (Bertsekas and Gallager, 1992).

$$W_k = \frac{\sum_{i=1}^{n} \lambda_i \overline{X}_i^2}{2(1-\rho_1-\ldots-\rho_{k-1})1-(1-\rho_1-\ldots-\rho_k)}.$$

As we might expect, priority queueing reduces delays for high-priority users at the expense of increased delays for low-priority users. However, priority queueing can also result in a situation in which one or more of the lower-priority classes of users receive no service at all. As

$$\sum_{i=1}^{k} \rho_i \to 1$$

from below, the mean waiting time for priority-$k$ customers (and for any customers with lower priorities) grows to infinity, i.e., the backlog of priority-$k$ customers waiting for service grows indefinitely with time.

The above results apply only for a system with no queue limit. For a system with a large but finite queue limit $m$, mean waiting times are always finite for all classes of customers. However, the quality of service for priority-$k$ customers still degrades severely as

$$\sum_{i=1}^{k} \rho_i \to 1;$$

mean waiting times tend to become very large (for those customers who are served) and blocking probabilities approach unity; reducing the queue limit reduces mean waiting time at the cost of increased blocking.

Although the above formulas apply only for the M/G/1 nonpreemptive priority case, it turns out that this type of behavior is characteristic of priority queueing systems in general unless some form of congestion management is used. *When the total arrival rate of requests of classes* k *and above exceeds the system capacity, users with priorities less than or equal to* k *will not be served.*

# ON DECEPTION

> Though fraud in other activities may be detestable, in the manage-
> ment of war it is laudable and glorious, and he who overcomes the
> enemy by fraud is as much to be praised as he who does so by force.
>
> —Machiavelli, *The Art of War*

We present in this appendix a brief survey of some of the relevant
concepts and terminology regarding the practice of battlefield deception, and posit its possible application to defensive information
operations. We strongly believe that well-crafted deception can play
a critical role as a protective measure; it is one of the 13 categories of
security techniques on which we focused in Chapters Four and Five.
Moreover, with few exceptions (e.g., Cohen & Associates' Deception
ToolKit), there currently seem to be inadequate attention and resources devoted to this topic within the "information survivability"
and computer security R&D communities. As aptly put by John
Woodward of MITRE (1997),

> It is interesting to note that the military has a long history of em-
> ploying deception in its warfighting, but has not yet embraced de-
> ception in its information systems, though these systems are touted
> as the battleground of the future.

Well-applied deceptions have aided combatants in both offense and
defense for the length of recorded history and the breadth of conflict,
from insurgency to invasion. In this discussion, we first provide general information about deception, then relate those concepts to information infrastructure security.

So what then is deception? Whaley (1982) has defined deception as

> information designed to manipulate the behavior of others by inducing them to accept a false or distorted presentation of their environment—physical, social, or political.

Deception is a proactive, purposeful enterprise; it is not the result of chance, nor the by-product of another endeavor, as noted by McCleskey (1991). The end-result of deception is a desired enemy behavior, which springs from a desired enemy mental state. The means of deception are used to produce this mental state, but the deceiver always bears in mind the ends toward which this artifice is being applied.

Successful deceptions rely heavily upon the accurate depiction and manipulation of the enemy decisionmaker's thought process. We may immediately wonder about the utility of defensive deceptions, fabricated without a particular adversary in mind. Reassuringly, the literature of cognitive and social psychology (the bedrock upon which deception theory rests) provides ample evidence of historical and cross-cultural validity for such efforts. As Lambert (1987) and Farnham (1988) point out, individuals process information with surprisingly consistent heuristics and are thus prone to similar logical and perceptual errors. For example, the "crying wolf" ploy, which conditions the adversary to a pattern of behavior as a method of reducing their readiness, is ubiquitous in the history of conflict, along both temporal and cultural axes. Well-constructed deceptions rely upon common human fallacies in information processing, in addition to intimate knowledge of a particular enemy. We may therefore conclude that deceptions, even without a particular adversary in mind, may prove quite useful.

## DECEPTION IN WARFARE GENERALLY

Deception is the alchemy of war: It transforms substance into shadow and vice versa. Deception is often thought of as simply camouflage, or disseminated falsehoods, or feints. But deception is more than just assets or practices, it is a planning process that combines operational flexibility with subterfuge to manipulate and exploit the behavior of the enemy.

The practice of deception is an interactive procedure, taking shape both from friendly as well as enemy assets, intentions, dispositions, and personnel. The deceiver works backward from the objective, some desired end-state in enemy behavior, asking "What is the response sought from the enemy?" Subsequently, the deceiver composes "the story," the beliefs that the enemy must hold in order to elicit the desired response. But if "the story" is the message, what shall be the medium? The deceiver must decide by what means "the story" will be told. And who is the audience? It is essential that the deception story be transmitted to the enemy decisionmaker who may order or take the desired action as set out in the objective. In relation to exploitation: How will the deceiver capitalize upon the successful deception? Is the target to be impeded, injured, or simply led astray? Let us consider each stage in more detail.

## Objective

The enemy response is composed of beliefs and reactions. Beliefs that do not precipitate reactions are practically worthless, and reactions will not occur without galvanizing beliefs. The following are some examples:

**Affecting Beliefs:**

- Establish a notion in enemy minds.

- Alter an existing enemy perception.

- Mask a friendly activity or attribute.

- Create confusion or "noise" to overwhelm enemy intelligence processing.

- Divert enemy attention.

- Condition the enemy to a pattern of friendly behavior.

**Engendering Reactions:**

- The enemy commits forces.

- The enemy withholds forces.

- The enemy mistimes actions.

- The enemy misplaces efforts.

## The Story

Working backwards from a desired end state, the deceiver asks, "What picture needs to be painted in order to elicit the response I seek from the enemy?" The story that is to be communicated emerges from the answer. Care must be taken to make the story both plausible and appropriate, given the situation at hand. Can what is being portrayed actually have occurred? Will it seem likely to the enemy? Moreover, the story must be communicated in enough detail, and with enough corroboration, that the enemy must take it seriously. Lastly, whatever enemy response the story seems to demand must be within the actual capabilities of the adversary (i.e., think twice before demanding mobility from a fixed opponent).

## The Means

Generally, deceptions have two broad types of components:

1. Ambiguity-based deceptions, whose aim is to raise the level of uncertainty in enemy minds.
2. Misdirection-based deceptions, whose aim is to make the enemy more certain, but of a falsehood.

Broken down further, deceptions usually combine one or more of the following elements:

- *Concealment or camouflage:* to prevent any detection of friendly forces, or to blend them in with the background noise such that they are not discerned.

- *Ruses and disguises:* to cloak belligerent forces in the appearance of nonhostile or even friendly forces.

- *Sensory saturation:* to overload the opponent's intelligence apparatus with a surfeit of information, creating a blanket of "noise" or simply paralysis due to the inability to process so much data.

- *Demonstrations and feints:* the conspicuous exhibition or application of assets to create either confusion or a misplaced certainty as to the whereabouts and nature of friendly activities.

- *Displays and decoys:* the conspicuous exhibition of real or phony assets in order to specifically draw attention or fire to that place and time.

- *Disinformation:* the transmission of distortions or patent falsehoods to the enemy intelligence apparatus.

An important note: The means of the deception must be tuned to the receiving capacity of the enemy. Clearly, it is of no use spinning an elaborate deception that relies critically upon phony radio transmissions if the enemy is not listening to the radio.

## Target

The deception story must be precisely aimed at the decisionmaker who bears the burden for the response. It is his or her particular perceptions and beliefs that form the foundation for any reactions on the enemy's part. If the story fails to reach the right target it may be virtually worthless.

## Exploitation

Often lost in the process of deception planning is any consideration of the leverage obtained. What will the deception, if successful, gain the deceiver? And crucially, how will that profit be spent?

## DECEPTION IN DEFENSIVE INFORMATION OPERATIONS

We consider here only defensive information operations and the manner in which deception may be employed to protect our information assets. As articulated by the Air Intelligence Agency's Air IW Center, deception may be used "to prevent an adversary's information gathering techniques" or other hostile IW actions. Considerations of offensive applications of deception, by either friendly or hostile forces, are beyond the scope of this present report.

Deceptive operations are interactive: We must identify our likely opponents, their disposition, their intent, their equipment, their methods, and their leadership. However, as noted earlier, deceptions may also rely upon common logical or perceptual errors in human information processing, as borne out by much cognitive study (Tversky and Kahneman, 1974; Kahneman and Tversky, 1983). The best-crafted deceptions utilize both instruments for maximum effectiveness, and for a wide range of applications (Dewar, 1989). Deception may in fact provide an outermost hedge against attackers, both by shaping the environment to avoid attacks altogether, and by proactively entrapping would-be adversaries.

We urge designers and administrators to consider some of the following objectives and outcomes when planning the application of defensive deceptions:

- Intruders are drawn into model systems (lures and ruses), where they may be observed, analyzed, and rendered harmless. This activity may also provide a basis for counterattacks.

- Intruders overlook access points, weak points, and critical resources through uniformity of appearance (camouflage) and deliberate mislabeling (disinformation).

- Intruders are dazzled with a superabundance of targets (sensory saturation) and are thus far less likely to strike at a critical resource in a restricted time frame.

- Intruders are presented with false access points and false files (decoys); willpower, time, and resources are wasted.

The exploitation of such deceptions may take many forms, but first and foremost they contribute to the survival and integrity of information assets.

# MAPPING SECURITY SOLUTION TECHNIQUES TO VULNERABILITIES

This appendix provides a detailed description of the matrix presented in Figure 5.1, which evaluates the applicability of the 13 protection technique categories introduced in Chapter Four to the 20 vulnerability attributes defined in Chapter Three. This mapping matrix is reproduced below, in Figure F.1, for ease of reference. The color in a particular cell of the matrix indicates our evaluation of the relationship between the associated protection technique category (column) and vulnerability attribute (row). There are five possible designations for these evaluations:[1]

- Addresses vulnerability (directly)—green

- Addresses vulnerability (indirectly)—light green

- Not applicable to vulnerability (directly)—blank

- May incur vulnerability (indirectly)—light yellow

- May incur vulnerability (directly)—yellow.

The remainder of the appendix is organized into 13 sections, one for each of the 13 security technique categories covered by the matrix. Each section includes a short paragraph for every vulnerability that, based on our evaluation, is affected by that security technique. These cell descriptions, which explain the reasoning behind the designations that we selected in each case, are grouped according to

---

[1]See Chapter Five for the definitions of these cell designations.

their designation and listed within each group in the order that they appear in Figure F.1.

## HETEROGENEITY

### Vulnerabilities Addressed (Directly)

*Homogeneity.* Heterogeneity reduces the vulnerability associated with homogeneity directly. The homogeneity of a system makes it more vulnerable because attacks employing techniques that work against one component or process of a system will also work equally well against other all components or processes of the same type. This creates the potential for a series of attacks, or a cascade of failures, that would result in widespread damage. Heterogeneity can greatly reduce the impact of attacks that take advantage of such vulnerabilities through differences in the system that limit the extent of cascading effects.

*Predictability.* Heterogeneity makes it more difficult for an attacker to predict the response of a system to an attack, thus making such attacks more difficult to design and conduct. Heterogeneity increases complexity because it generates a multiplicity of new interactions among different element types. This added complexity makes it more difficult for anyone, especially an attacker who is not thoroughly familiar with the system, to reliably predict the system's behavior. As a result, a potential attacker will find it more difficult to construct and implement a successful attack, since he or she cannot anticipate and capitalize as easily on the system's responses to his or her actions.

*Malleability.* Heterogeneity reduces malleability vulnerabilities by making it less likely that an attacker will have the knowledge, tools, and skills necessary to successfully manipulate all of the different types of elements in the system. The actual malleability of the system is not necessarily affected by heterogeneity, but the increased number of element types effectively reduces the vulnerability associated with any malleability that remains.

*Electromagnetic Susceptibility.* Heterogeneity can help reduce the electromagnetic susceptibility of a system by designing in differing levels of resistance or shielding to electromagnetic pulses. In addition, the use of multiple types of system components will reduce the effectiveness of various methods that attempt to "tap" or interfere with the electromagnetic emissions of computer hardware and communications lines. Greater diversity assures that no one method of electromagnetic attack will be effective throughout the system.

*Dependency (on Supporting Facilities/Infrastructure).* Dependency on supporting infrastructures can be reduced by adding heterogeneity into the set of infrastructures that the system relies upon. If the dependent system has multiple sources of infrastructure support—for example, multiple electric power sources such as a generator in addition to public utility power lines—then the vulnerability incurred by the dependency is diminished. In this way, if one supporting infrastructure fails, there is a different source that can take its place. It is important to note, however, that the backup infrastructures should not be subject to the same set of vulnerabilities as the primary infrastructure.

## Potential Vulnerabilities Incurred (Directly)

*Sensitivity.* The added complexity of heterogeneity may incur an unintended vulnerability by making the system more sensitive to small accidents or attacks. Complexity is cause for concern because it creates more opportunities for the system to "wander" into unforeseen circumstances, where it may exhibit unfamiliar behavior, thus making it more prone to failure, and possibly even sabotage.[2]

*Difficulty of Management.* The more difficult it is to correctly configure and maintain a system, the more likely it is that vulnerabilities will arise through oversight in security procedures. As the heterogeneity in a system increases, the management requirements to properly configure and maintain it also increase, and it becomes more likely that mistakes will be made. Such vulnerabilities are especially great if the level of effort and degree of expertise required to administer the heterogeneous system exceed the resources and capabilities of the system administrators.

---

[2]This argument draws on the ideas of Charles Perrow, 1984.

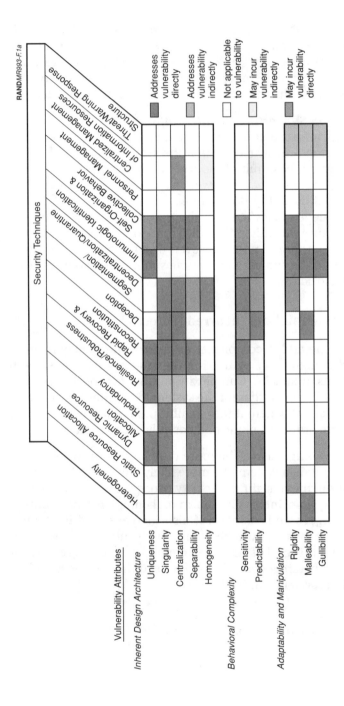

**Figure F.1—A Matrix Showing the Applicability of Security Techniques to Sources of Vulnerability**

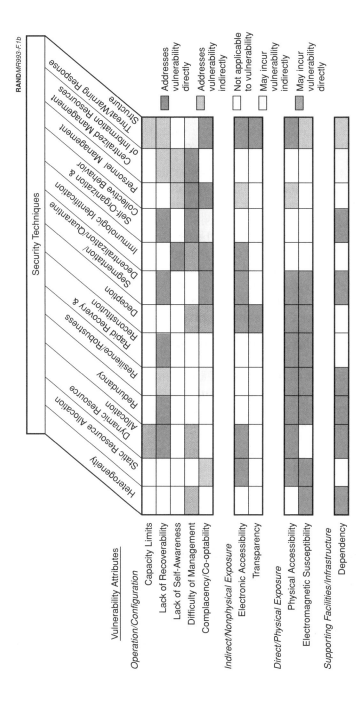

Figure F.1—continued

## STATIC RESOURCE ALLOCATION (SRA)

### Vulnerabilities Addressed (Directly)

*Singularity.* SRA addresses singularity vulnerabilities directly by always devoting more security resources to singular elements of a system that are critically important. Even though this extra protection does not eliminate the singularity, it does significantly reduce the likelihood that it will be successfully exploited by an attacker.

*Electronic Accessibility.* SRA reduces this source vulnerability by limiting electronic access, where possible, to critical elements of the system, while also providing those elements with more protection. SRA can address electronic accessibility directly by limiting or restricting access to those parts of the system that are determined, a priori, to be the most critical or important. This approach, however, may not always be available or affordable. Indirect SRA approaches provide alternatives that do not reduce accessibility, but instead provide the critical parts of the system with additional security resources that reduce the likelihood and impact of a successful attack.

*Physical Accessibility.* SRA addresses physical accessibility directly by using barriers—such as locked doors, guards, fences, and walls—to protect the critical elements of a system. Although this will not deny access to everyone, it does limit access to a smaller, more select group of individuals, thereby reducing the likelihood of an attacker's taking advantage of physical accessibility.

*Electromagnetic Susceptibility.* SRA includes the concept of selective hardening. In this instance, the shielding of system components that may be vulnerable to electromagnetic pulses, or other electromagnetic methods, can reduce the threat of electromagnetic susceptibility.

### Vulnerabilities Addressed (Indirectly)

*Complacency/Co-optability.* SRA can have an indirect effect of reducing the problems associated with complacency if it includes a standard list of requirements, such as a procedure checklist, that would have to be checked and satisfied on a regular basis. A static list of priorities, or a "to do" list, might increase the probability that administrative personnel would follow proper security procedures. This is only an indirect effect, however, since personnel could still choose to not follow these procedures, even though they are structured to make such noncompliance more difficult.

## Potential Vulnerabilities Incurred (Directly)

*Separability.* SRA identifies and then protects the critical nodes and links of a system; but, by focusing on specific pieces of the system, SRA can create new separation vulnerabilities. For example, SRA may limit the number of pathways into and out of a critical node. That, however, may make it easier for an attacker to separate and besiege that node by cutting off or disabling those few links that connect it to the rest of the system. Separability can thus negate some of the advantages of SRA by enabling an attacker to isolate and deny a portion of the system without having to penetrate its hardened defenses.

*Rigidity.* A static technique can cause the system to become more rigid, which can induce certain types of vulnerabilities. Because SRA determines where to position all of the security resources a priori, the system is unable to adapt its protective posture to cope with unforeseen circumstances and unanticipated new threats. Also, by "playing first" in selecting a static allocation of resources, the defender gives the attacker an implicit advantage; the attacker can observe the nature and location of the fixed defenses and then optimize his or her attack plan to avoid, deceive, and even take advantage of them.

## Potential Vulnerabilities Incurred (Indirectly)

*Centralization.* The use of SRA to protect a system can generate a tendency toward greater centralization to take advantage of fixed defenses. There is a temptation to put more "eggs in one basket" by locating more important elements inside the fortified "city walls." New centralization vulnerabilities can emerge as a result of this subtle effect, especially if security is lacking inside the areas with extra protection (e.g., "trusted" information systems).

# DYNAMIC RESOURCE ALLOCATION (DRA)

## Vulnerabilities Addressed (Directly)

*Uniqueness.* DRA allows the system to respond flexibly to unforeseen problems caused by unique elements that have not been thoroughly tested and debugged. The vulnerabilities associated with one-of-a-kind components and processes are difficult to anticipate, so a dynamic approach to protection is better able to respond to the problems they create than a static one. In addition, a DRA approach can be adapted and improved over time as experience accumulates regarding the vulnerabilities associated with uniqueness.

*Singularity.* The dynamic and comprehensive nature of DRA makes it better suited for robust, systemwide protection rather than the preselected, fixed, local defenses of SRA. In addition, most DRA approaches can be designed to provide extra protection to critical singularities. Thus, singularity vulnerabilities can be reduced by DRA, although SRA would probably address them more efficiently.

*Separability.* Shifting from SRA to DRA will eliminate the potential for increased vulnerability to separation, since it is more difficult for an attacker to separate and isolate an important portion of the system when security resources are mobile rather than fixed. In fact, DRA can reduce the risk of separation anywhere in the system during an attack, independent of any static defenses, by shifting resources to those parts of the system that are in danger of being disconnected.

*Predictability.* DRA is able to reduce vulnerabilities associated with predictability because of its dynamic and responsive nature. A potential attacker will have more difficulty predicting where he or she will encounter increased resistance during an attack when DRA has been implemented. It follows that DRA is less effective when its responses can be easily anticipated by attackers, so continual adaptation and innovation is essential to the success of DRA in reducing predictability.

*Lack of Recoverability.* DRA can enable a system to recover more quickly by shifting resources to wherever they are needed during and after an attack. In particular, DRA can help limit damage and retain functionality during an attack and maybe even enable the system to optimize its performance while in a degraded state after an attack.

Such capabilities would be especially beneficial in a system that is unable to repair itself in a timely manner.

*Electronic Accessibility.* Like SRA, DRA can reduce accessibility vulnerabilities by either restricting access or improving security in certain portions of the system. Unlike SRA, however, DRA is not limited to a predetermined posture; it can combine access restrictions and heightened security in a flexible and responsive manner throughout the system. In this sense, DRA is a more robust way of reducing accessibility vulnerabilities than SRA is.

*Physical Accessibility.* By dynamically reallocating resources to unaffected portions of a system after a physical breach or disruption, the functionality of the system may be retained, or at least remain less degraded. Although DRA does not directly address the physical accessibility problem, it can react to an attack and assist in preventing further damage to the system, thereby reducing the consequences of this vulnerability.

*Dependency (on Supporting Facilities/Infrastructure).* The vulnerability associated with dependency can be reduced by using a planned set of rules to reallocate the demand for a particular  resource, such as phone lines, that is supplied by one or more supporting infrastructures. This may take the form of automatically switching to a dedicated backup source if the primary source is lost or damaged, or reallocating services among the set of other infrastructure sources that are still available.

## Potential Vulnerabilities Incurred (Directly)

*Sensitivity.* The use of DRA could introduce new vulnerabilities associated with sensitivity. In particular, an attacker might, with sufficient knowledge and skill, be able to use inherent sensitivities to intentionally overstimulate the DRA capability of the system, and thereby improve the likelihood, and even the severity, of a successful attack.

*Gullibility.* DRA can increase the gullibility of the system and thus make it more vulnerable to certain types of attack. An attacker could, for example, create a diversion that would induce the DRA capability to respond in a manner that he or she could then take advantage of.

There might also be specific gullibilities in a particular DRA approach that a smart or experienced attacker could exploit.

*Capacity Limits.* There may be some danger that a DRA capability could inadvertently demand too much of its host system. Indeed, an attacker might be able to intentionally cause a DRA capability to saturate itself, or even the entire system, in order to disrupt or impede the normal functioning of the system. It should be noted that the extent and importance of these vulnerabilities depend greatly on the size, nature, and purpose of the system involved.

*Difficulty of Management.* The use of DRA to improve the security of a system will necessarily make that system more difficult to manage. The DRA component of the system will need to be configured properly, which could require considerable effort and expertise. In addition, DRA will affect various components and processes within the system. Unless these interactions are actively and sensibly managed, they may create holes in the protection provided by DRA techniques.

## REDUNDANCY

### Vulnerabilities Addressed (Directly)

*Singularity.* Redundancy can assist in reducing vulnerabilities associated with singularity in two ways: (1) maintaining spare parts, information, or components for use in the repair or replacement of damaged or corrupted singular system elements; and (2) creating one or more whole duplicates of the singular element, thus making it no longer a singularity. Since a singular process or component acts as a lightning rod for attacks, retaining one or more backups helps mitigate the risk. The singularity may still draw attacks to it, but with the additional components or information, the damaged or corrupted portions of the singular element can be replaced. Creating more than one of the element would, of course, largely negate the lightning rod effect.

*Separability.* If a system or process can be isolated in a "divide and conquer" strategy, then introducing more redundancy into the system will reduce this vulnerability. The attacker may still try the same strategy but will have to isolate more systems before being able to defeat them. If the system is singular, maintaining redundant "spare parts" may not help to mitigate the danger of isolation; rather a whole extra system would be needed.

*Lack of Recoverability.* Redundancy reduces vulnerabilities associated with a lack of recoverability in two ways: (1) maintaining redundant components or information extends the life of a system by allowing replacement of damaged or corrupted components; and (2) creating entire duplicates of various system elements means that the attacked portions of the system can perish and remain unrecoverable while system function is still restorable.

*Physical Accessibility.* Redundancy can reduce vulnerabilities associated with physical accessibility by providing for multiple systems or components to take over for those elements that are damaged or compromised by a physical attack. The physical accessibility problem would still exist, but by having redundant components and parts, the processes and capabilities of the system would be able to continue, and the impact of the attack would be greatly diminished.

*Electromagnetic Susceptibility.* Just as in the case of physical accessibility, redundancy addresses this vulnerability by keeping multiple system components available for use as backups in the event of an electromagnetic attack. These systems or components may still be vulnerable to electromagnetic pulses and other attack methods, but even if some are lost, others would be able to take over their functions and processes.

*Dependency (on Supporting Facilities/Infrastructure).* Having multiple sources of supporting infrastructure (e.g., more than one phone line) lessens the system's dependency upon a single infrastructure. If one of the sources fails, another can take its place.

## Potential Vulnerabilities Incurred (Directly)

*Homogeneity.* Once an attacker discovers a weakness in a homogeneous system, the same attack can be used again and again, on each of the homogeneous system elements. Redundancy perpetuates this problem, rather than solving it. Although redundant "spare parts" might be able to bring the system back to life, the vulnerability to the attack remains.

## RESILIENCE AND ROBUSTNESS

### Vulnerabilities Addressed (Directly)

*Uniqueness.* A unique system may have multiple unexpected features (i.e., bugs) that could lead to increased vulnerability. One resilience and robustness technique is to test and model systems to find and correct these unexpected features, and thus reduce the vulnerability of the system. Although modeling and testing may not find all of the bugs, it should make the system considerably more resistant to attack.

*Physical Accessibility.* Resilience and robustness can help reduce vulnerabilities associated with physical accessibility by making the system and its components more resistant to damage. Portions of the system may be physically accessible, but if they are more resilient or more robust to a physical attack, then the level of damage that they experience will be reduced. Resilience and robustness techniques do not remove this vulnerability, rather they make the potential damage associated with it less severe.

*Electromagnetic Susceptibility.* By making system components more resilient to electromagnetic pulses, or more robust to electromagnetic interception and interference methods, the system as a whole becomes less vulnerable to this form of attack. The components may not be impervious to this type of attack, but the impact of the associated vulnerability is reduced.

*Dependency (on Supporting Facilities/Infrastructure).* Dependency may be reduced by enabling a system to withstand drops or losses in its external supply of a particular resource from a supporting infrastructure. An increase in the tolerance of a system and its various elements to faults and failures also helps reduce its dependency vulnerabilities.

### Vulnerabilities Addressed (Indirectly)

*Singularity.* If a component or process exists in only a single place, then making that component more robust or more resilient to an attack will reduce its vulnerability. This approach does not, however, divert the attention of a potential attacker from the singular component, nor does it eliminate the singularity itself, so the effects of such techniques on singularity vulnerabilities will tend to be fairly mild, unless combined with other more direct approaches.

*Centralization.* Resilience and robustness improve the ability of a centralized system to resist and endure an attack. Those portions of a system that are highly centralized are, like a singular system component, prime targets for an attack. Ensuring that a centralized system can absorb or withstand an attack does not make it that much less attractive as a target and does not directly address its centralization vulnerability. However, it does reduce the impact of the attack to some extent.

*Homogeneity.* Homogeneity is another vulnerability that is only indirectly addressed by resilience and robustness because resistance is increased while the inherent vulnerability remains. Once a security flaw in a homogeneous system is discovered, the same flaw can be exploited by an attacker throughout the system, regardless of how robust or resilient the system is. Making a homogeneous system more robust and resilient is merely a partial solution to this intrinsic source of vulnerability.

*Sensitivity.* Resilience and robustness reduce this vulnerability by enabling the critical components and processes to withstand an attack that takes advantage of sensitivity. Making various parts of a system more resilient and robust does not make it less sensitive; a small stimulus can still cause a large, complex information system to behave in a dangerous or unpredictable manner, even if its components are robust and resilient. But making the system tougher may help it to survive and withstand a sensitivity-based attack, thus somewhat reducing the impact of such an attack.

*Lack of Recoverability.* Resilience and robustness do not resolve the vulnerability associated with a lack of recoverability. A system with more resilient and robust components is, however, better able to resist and withstand an attack and as a result will experience fewer large failures due to an attack. If the system can resist almost any attack, then its lack of recoverability is less of a concern, even though that vulnerability is still present.

## Potential Vulnerabilities Incurred (Indirectly)

*Complacency/Co-optability.* Resilience and robustness can give system users and administrators a false sense of security. This misperception can induce them to become complacent, and thus less diligent in applying existing security procedures. As they realize that the system can better withstand potential attacks as a result of its resilience and robustness, they may reduce the level of vigilance in their day-to-day operations, and thereby make the system more vulnerable to attack. System managers may also underestimate the overall vulnerability of their system. They may think that resilience and robustness by themselves are enough to withstand or absorb (and possibly even deter) most attacks, even though other security measures could be applied that would address the vulnerabilities of their system more directly.

## RAPID RECOVERY AND RECONSTITUTION

### Vulnerabilities Addressed (Directly)

*Uniqueness.* The vulnerability of a unique system comes from the fact that it has not been thoroughly tested and so may have some unknown weaknesses and vulnerabilities. Yet, if such a system can easily be reconstituted after an attack, the vulnerability is less severe because the overall impact is not so significant. The uniqueness vulnerabilities will still exist, but the outcome from the loss of the system can be greatly reduced if the system can quickly come back to life.

*Singularity.* Rapid recovery and reconstitution reduce the vulnerability of singularity by ensuring that the singular system element has the ability to recover after being attacked. If an attacked system is able to quickly "bounce back" from the assault, then the attack will be of less value to the attacker. Although possessing the ability to rapidly recover or reconstitute itself does not solve the singular nature of the system, it does assist in preserving the system's functionality after an attack.

*Centralization.* Rapid recovery and reconstitution reduce the vulnerability of centralization by ensuring the system can rebound after undergoing an attack. A centralized system creates a vulnerability since other systems or functions are dependent on the central entity. Providing a way for the system to swiftly resume its role decreases the vulnerability. Again, the recovery and reconstitution mechanisms do not eliminate the vulnerability; rather they merely reduce its ramifications.

*Separability.* Rapid recovery and reconstitution may reduce the vulnerability of separability by rapidly restoring the damaged portion of the system as well as the links between it and the rest of the system. If, however, only the system is restored, and the links between itself and the outside network or system are still severed, then the vulnerability will remain. Also, the ability to recover quickly from a separation does not eliminate the original vulnerability that caused the separation in the first place. Even so, if the recovery and reconstitution are truly rapid, then these concerns will matter very little since the impact of a separation attack will be minimal.

*Sensitivity.* Rapid recovery and reconstitution diminish the vulnerability of sensitivity by helping the sensitive system quickly resume its operations after an attack. A sensitive system can easily be affected by a malicious attack, but if it can also easily recover and reconstitute to resume its normal functions, then the severity of the attack is greatly mitigated. The vulnerability itself is not removed; rather, its consequences are significantly diminished.

*Lack of Recoverability.* The very nature of this vulnerability is addressed by rapid recovery and reconstitution; a system and its components are designed and configured so that they can recover quickly from an attack or failure to resume their normal functions.

*Physical Accessibility.* Although important parts of a system may be physically accessible, if those elements have the capability to rapidly recover from a physical attack, then the related vulnerability is greatly reduced. By having the ability to "spring back" after an attack, the critical functions performed by the system may be resumed almost immediately after an attack, such that no significant degradation is experienced.

*Electromagnetic Susceptibility.* Similarly, a system that can rapidly recover from an electromagnetic attack is less vulnerable to such methods of attack. The system may still be susceptible to an electromagnetic attack, but any damage caused by such an attack will not be severe or long-lasting.

## DECEPTION

### Vulnerabilities Addressed (Directly)

*Singularity.* Any identifiable choke point in a system would benefit from the application of deception. Historically, very few defenses have turned out to be impregnable, so deception may be able to provide an additional bulwark against an attack. Specifically, the location, character, or even existence of a critical singularity could be concealed, camouflaged among innocuous associates, or disguised to appear as something else entirely. False information about the nature and location of singularities in a system could also be disseminated, with the goal of leading astray, or even entrapping, potential attackers.

*Centralization.* When topology becomes a liability, deceptions can offer an effective, although somewhat indirect, remedy. Just as singularity vulnerabilities can be addressed by deception, centralized portions of a system can also be hidden completely, buried in "noise," or intentionally mislabeled, and decoys that resemble centralized system elements can mislead or even trap an opportunistic attacker. A series of baffling deceptions can sap the willpower of even the most ardent infiltrator, even though they have no impact on the underlying source of centralization vulnerabilities.

*Predictability.* Deceptions may be used to introduce irregularity into a predictable system by concealing certain aspects of function or form. Imagine the system in question is a handgun, and every one knows that pulling the trigger will cause the hammer to fall and the gun to fire. Without affecting the predictability, we can hinder our opponent by making the trigger look like something else (e.g., the magazine ejection button). We may leave the original trigger inert or perhaps wire it to lock the safety in the "on" position when it is touched. Now, intruders will either waste time or lock the gun against themselves, while anyone privy to the deception can still fire the gun handily by manipulating the new trigger.

*Malleability.* If a system allows its users, including those who have gained unauthorized access, somewhat free reign once they are inside the system, then deceptions may be able to hinder them from wreaking too much havoc. Critical files may be intentionally mislabeled, essential capabilities may be hidden or misadvertised, and enticing false targets may be used to lure an attacker into a trap or cause him or her to trip an alarm. Deceptive approaches like these can confuse or mislead an attacker, making it more difficult for him or her to take advantage of any existing malleability weaknesses, even though those vulnerabilities are not actually removed.

*Electronic Accessibility.* Even if a system cannot be hermetically sealed or vigilantly guarded, implementing deceptions may well thwart the intentions of an attacker. Actual access points may be concealed or camouflaged; false access points may divert or ensnare attackers; false information about access procedures may be disseminated; system activities may be hidden, such that access attempts appear to be ineffectual to those not privy to the deception. Even if deceptions are ultimately penetrated, the delays and distractions that they create may be sufficient to erode the attacker's determination to continue.

*Transparency.* Some of the most effective deceptions take place without a trace of opacity. Decoys and diversions, sensory saturation, and disguises are all excellent examples of transparent deceptions. For example, critical files may be mislabeled, buried amidst an ocean of other mundane information, or hidden using phony stand-ins. Deception turns the tables on most transparency vulnerabilities, taking advantage of them to manipulate, distract, mislead, or confuse potential attackers.

*Physical Accessibility.* Deception may reduce this vulnerability by confusing the attacker, or pointing him or her in the wrong direction, away from physically accessible system components. Deceptions will not eliminate accessibility to the system, but they will cause any potential attacker to have a more difficult time recognizing real targets, and as a result will force the attacker to use more resources to mount a successful attack.

*Electromagnetic Susceptibility.* Like the use of deception to mitigate physical accessibility, deception ploys will not remove this system attribute, but can significantly reduce the vulnerability associated with it by confusing or misdirecting an adversary's attack. While still susceptible to electromagnetic attacks of various types, the system or component may be well enough hidden through deception that an attacker will be unable to properly direct his or her attack.

## Potential Vulnerabilities Incurred (Directly)

*Sensitivity.* Deceptions add to the complexity, and thus effort, of any operational plan or system. The "monkey's paw" maxim applies here: beware of unintended, unpleasant side effects. The more complex a deception becomes, the more likely it is that some unforeseen factor or event will interfere with the deception. Such complications are, at best, merely annoying and inconvenient, but they can easily undermine the entire plan, and can possibly even backfire on the deceivers.

*Difficulty of Management.* Deception may increase a system's vulnerability by making the system increasingly difficult to manage and thereby raising the potential for security mistakes or oversights. Deception planners must also determine who needs to know about the deception and ensure that those people who are involved are able to avoid or bypass the deception readily, without revealing it through their actions. It requires a delicate balance between having enough people involved, avoiding problems with legitimate but unwary system users, and effectively manipulating an attacker's perceptions. It requires both deftness and minimalism to keep the management of the deception as simple as possible, both in terms of technical configuration and personal interactions.

*Complacency/Co-optability.* Trust, discipline, and secrecy are essential to the success of any deception plan. Thus, the vulnerabilities associated with complacency and co-optability increase when deception is employed to protect a system. Leaks, both intentional and accidental, can easily undermine or even destroy a deception that would otherwise have been very effective. The wider the circle of people with knowledge of the deception, the more likely it is that at least one of them will either betray the trust placed in him or her, for any number of reasons, or through complacency reveal some aspect of the deception to a potential attacker. Indeed, the phrase "loose lips sink ships" also applies to information system security.

## SEGMENTATION, DECENTRALIZATION, AND QUARANTINE (S/D/Q)

### Vulnerabilities Addressed (Directly)

*Singularity.* Decentralization addresses singularity vulnerabilities directly by distributing the functions and processes of a singular system element to various other parts of the system. In addition, segmentation and quarantine can prevent damage already incurred elsewhere in the system from spreading to a critical singularity.

*Centralization.* Just as in the case of singularity, decentralization can directly reduce the source of this vulnerability, while segmentation and quarantine can act to prevent the propagation of damage into important centralized portions of a system.[3]

*Homogeneity.* A classic epidemiological case of this type of protection is the use of quarantine to prevent a virus from spreading unchecked through a homogeneous population. Similarly for an information system, S/D/Q can prevent an attacker from using the homogeneity of a system to bring it down entirely. Even so, this approach does not directly influence the homogeneity of the system; rather, it greatly limits the magnitude of the vulnerability associated with this attribute.

*Sensitivity.* S/D/Q could prevent small perturbations in a system— due to an attack, an accident, or just unusual circumstances—from leading to catastrophic failures and subsequent loss of function. Such sensitivity vulnerabilities could be reduced by (1) "sealing off" any stimuli that are observed to be interacting in a dangerous manner, (2) segmenting the system to limit the spread of damage, or (3) decentralizing functions to restrict the loss of functionality due to damage in a single area. These changes may also make the system less complex, and hence less susceptible to sensitivity problems in the first place.

*Lack of Recoverability.* While S/D/Q does not help restore a system to its normal state after an attack, it does limit the degree of degrada-

---

[3]If the damage or intrusion actually *begins* in a critical part of the system, then S/D/Q is of dubious value.

tion that the system will experience. Thus, S/D/Q techniques tend to reduce the potential impact of the vulnerabilities associated with poor recoverability.

*Complacency/Co-optability.* S/D/Q reduces the vulnerability associated with co-optability in two ways. First, the damage caused by a co-opted individual may be confined to a limited area through quarantine and segmentation. Second, the decentralization of important system functions makes it more difficult for a single renegade insider to disable the entire system. S/D/Q does not negate the possibility of a trusted individual being co-opted, but it can significantly reduce the ability of a co-opted individual to cause widespread damage.

*Electronic Accessibility.* Even if a system suffers from excessive openness, S/D/Q prevents the damage wrought by an electronic intruder from being catastrophic or pervasive. Moreover, the attacker may be sealed inside the damaged segment or quarantined area, and thus unable to continue his or her attack, or cause further damage.

*Physical Accessibility.* Although S/D/Q does not keep an adversary from physically accessing a system or its components, it does reduce the vulnerability by limiting the extent of the potential damage through segmentation and quarantine. As the system is being physically attacked, it can close off physical access to other parts of the system to reduce the overall level of damage. Also, decentralization will permit the system to continue functioning even if certain portions of it are physically attacked.

*Electromagnetic Susceptibility.* By segmenting and decentralizing the system, the amount of damage that would be caused by an electromagnetic pulse attack is limited. Parts of the system may indeed be damaged by the attack, but if the system is not forced to rely on each individual component, the functions of the system may continue.

*Dependency (on Supporting Facilities/Infrastructure).* S/D/Q can be used to directly reduce a system's dependency on supporting facilities and infrastructure. In the event of a loss or failure of some part or aspect of the supporting infrastructure, only certain portions of the system would be affected rather than the entire system. Likewise, infrastructure and facilities could be quarantined to isolate any effects that might spread to other parts of the system. Decentralization is particularly helpful in reducing dependency. For example, functions that can be performed in several locations are retained even when critical infrastructures, such as electric power, are lost in one location.

## Potential Vulnerabilities Incurred (Directly)

*Separability.* Practically by definition, if an attacker can coerce a system into implementing a quarantine or segmenting itself, it has become divided and can subsequently be isolated for attack. If the system relies on synergy—where the system elements are weaker on their own than as part of a unified system—then its division will provide an attacker with a significant force multiplier. Such a reversal of synergy as a result of separation could arise from resource constraints: Resources that are usually available to all parts of the system may be limited to a particular segment and thus unavailable to other parts.

*Predictability.* A clever attacker can make use of S/D/Q implementations to deny service by "pushing the right buttons" that will trigger a segmented state, in which services or performance are presumably reduced. Furthermore, a quarantine by definition dooms everything in the quarantined area to suffer the damaging effects of the attack, *even if they have not already been exposed.* This could increase the ability of an attacker to wreak havoc at least locally, since he or she can predict that the quarantined area will be abandoned, allowing him or her to operate uncontested within that area.

*Rigidity.* Since S/D/Q is a damage-containment feature, and not a fighting feature, there must be thresholds of damage or deleterious effects that will trigger the sealing-off response. That said, to protect the system as a whole from mortal harm, strict adherence to such rules must be in effect. This response, however, is a rigid one, as it impedes any attempt to react to an attack in a more flexible fashion. For example, a system could bide time to more closely observe an intruder's behavior, even though this may entail suffering substantial local damage, since it would enable the rest of the system to be protected more effectively.

## IMMUNOLOGIC IDENTIFICATION

### Vulnerabilities Addressed (Directly)

*Uniqueness.* In a setting where unforeseen security bugs may derive from the novelty or solitary nature of a system, immunologic identification techniques provide a hedge against the exploitation of such vulnerabilities through the techniques' adaptiveness and broad preparedness. Such defensive measures are ever vigilant and omnipresent, and they learn from their experiences, providing an excellent means of containing or altogether preventing any exploitation of "unexpected features."

*Predictability.* Immunologic identification serves as a hedge against inflexibility. It can prevent an attacker from being able to predict, and then prepare for, the system's response to his or her intrusion and subsequent damaging actions. This is accomplished with the probabilistic method of partial matching, which undercuts the attacker's calculations.

*Rigidity.* Immunologic identification is an excellent prescription for avoiding the exploitation of this vulnerability. If a system is designed without the capability to react quickly to intrusion or subversion, some flexibility and dynamism in real-time defense can be added or built by implementing an immunologic identification process. Moreover, through systemwide communication and the capacity to learn from experience, this type of approach can enable the system to be better prepared for future conflicts, in spite of any inherent rigidity.

*Malleability.* This vulnerability concerns the interiorization of antagonists. In this case an attacker's ability to wreak havoc due to a particularly pliant operating environment is addressed. Immunologic identification solves this problem through ubiquitous vigilance, which is able to recognize irregular behavior and thus determine *when malleability is being abused*, and then trigger an appropriate alarm or countermeasure.

*Gullibility.* The adaptive, vigilant nature of immunologic identification makes it particularly resistant to deceptions. With a continuous, skeptical (not to say jaundiced) eye, this approach guards all parts of the system, at all times. No user or process is exempt from interrogation, so if anomalous actions occur at any point in the access or use of system resources, the immunologic identification implementation will trigger an appropriate response, even if the person or system in question is supposed to be "trusted." Moreover, the detection algorithms used in this approach are sensitive—and may be augmented by experience or instruction—further increasing the system's resistance to deception.

*Lack of Self-Awareness.* Immunological identification is an excellent prescription for a system that has no introspective qualities, since it can act in some sense as a "sapience graft." Learning, cooperation, flexibility, and readiness—all qualities exemplified by the immunologic identification principle—are hallmarks of self-awareness.

*Complacency/Co-optability.* For the reasons enumerated above under malleability, gullibility, and lack of self-awareness, immunologic identification is clearly applicable to this vulnerability as well. Users—whether unfriendly from the outset or recrudescent—are always under the eye of a security system based on immunologic identification. All actions are continuously monitored and periodically interrogated to ensure fidelity. This means that even trusted individuals who have been co-opted would still be faced with a security system that always questions and analyzes their actions. As a result, there is no "safe haven" for an attacker or co-opted insider, since no location or individual is exempt from being scrutinized.

*Electronic Accessibility.* Systems with a great deal of openness absolutely require a broad defensive front: They are likely to suffer a range of attacks that vary in character, space, and time. The plastic, continuous, omnipresent defense offered by immunologic identification meets that need. Electronic accessibility implies that it is difficult to prevent a determined attacker from getting in, but immunologic identification watches all interior structures and behaviors for dangerous foreign or abnormal actions, which it responds to by triggering alarms or countermeasures as appropriate.

## Potential Vulnerabilities Incurred (Directly)

*Sensitivity.* The wide range of non-self detections (so useful in the identification of novel threats), coupled with the autonomous function of immune agents, may lead to significant overhead and numerous false alarms. Worst of all, legitimate users engaging in nonhostile action may inadvertently trigger alarms or countermeasures; clever, hostile intruders may also find ways to trigger such autoimmune reactions against legitimate users. These sensitivity problems all stem from the added complexity associated with integrating immunological identification into an already complex system. A classic example might be the neural net processes used today to recognize anomalous credit card usage. This is good if the credit card has actually been stolen, but would cause problems otherwise. If the transaction is impeded, it is a mere hindrance; but if the card is revoked, then the situation is a more serious case of service denial.

*Difficulty of Management.* Any system that implements some form of immunological identification as a security measure will inevitably become more difficult to manage, thereby increasing the potential for mistakes and oversights. The system will need to be configured correctly, which may be especially difficult initially when the new security features are being integrated into the system. The "immune response" of the system will also need to be monitored and refined constantly, so as to balance security requirements with the efficiency and convenience concerns of legitimate system users. Problems associated with meeting these management challenges will undoubtedly create some unintended flaws or weaknesses, which could conceivably be exploited by a knowledgeable and experienced attacker.

## SELF-ORGANIZATION AND COLLECTIVE BEHAVIOR[4]

### Vulnerabilities Addressed (Directly)

*Singularity.* The collective resilience and dynamism of autonomous agents, which shed and shift loads in response to changes in the operating environment, practically remove singular "choke points" from a system. When switching stations or connections fail, the system tends to reroute rather than simply fail, and when a particular

---

[4]Note that many of the hypothesized advantages thought to be conferred by this category of protection techniques are, at this time, still unproved. Thus, the evaluations here are based largely on speculation rather than on specific successful examples. Nonetheless, enough demonstrative and provocative work has been done on the subject to warrant its inclusion in our analysis.

node fails, its duties are reassigned and seamlessly picked up by others.

*Centralization.* By its very definition, the collective behavior associated with self-organization stems from decentralized control. The devolution of authority to autonomous agents precludes the possibility that a centralized part of the system would, if disabled, threaten the functioning of the system as a whole. The system would simply reorganize itself to compensate for the damage, creating new regions of centralization where appropriate.

*Separability.* The ubiquity and adaptability of autonomous agents tend to undermine most "divide and conquer" strategies, limiting their chances of success. The system will continue to function normally even if some parts of it are isolated. Moreover, the dynamism of this security approach implies that attempts at separation will meet with defensive rerouting.

*Rigidity.* Not only will this methodology establish "best" procedures based upon initial conditions, but as the environment evolves, optimization, goal-seeking, and adaptation will continue. As such, this protective technique can be seen as a cure for rigidity, conferring adaptability and resilience through collective and goal-seeking behavior. This added flexibility enables more effective responses to attacks.

*Lack of Recoverability.* Even in a degraded state, local optimization continues as a result of self-organization. Moreover, load-shifting and load-shedding among collections of agents imbue a system with a greatly improved ability to recover and heal itself. These techniques allow the system to degrade gracefully and possibly with some self-healing capability, thus reducing the impact of various types of attacks.

*Difficulty of Management.* The principle of self-organization implies that the necessity for oversight is reduced significantly, if not altogether eliminated. Note that this is not self-policing, but merely self-managing. This reduces the likelihood of management errors and oversights that lead to security holes.

*Dependency.* Though not able to wean assets off their supporting substrates, self-organizing optimization and the resilience conferred by autonomous agents may offer enhanced survivability. If supporting infrastructures are damaged or compromised, rather than halt or fail, systems applying self-organizing and collective behavior will tend to optimize at reduced or degraded states. They will thus retain functionality, at least until the point where a critical underpinning is completely knocked out.

## Potential Vulnerabilities Incurred (Directly)

*Uniqueness.* Much of the advantage to be gained from this security approach is from "emergent" properties that are not directly built or designed into the system. Concomitant with the potential benefits of such approaches are some important risks associated with unexpected behavior and phenomena, which emerge only after the system has been implemented and is in use. An attacker could observe these unanticipated features and learn to take advantage of them.

*Sensitivity.* The complexity and pervasive influence of self-organizing and collective behavior may be significant. As a result, it is possible that relatively small perturbations of the system or its environment may lead to unexpectedly large and damaging responses. A clever attacker could seek to take advantage of such sensitivity by continually stimulating the system, hoping to hit a sensitive spot and initiate a large failure or accident.

## Potential Vulnerabilities Incurred (Indirectly)

*Complacency/Co-optability.* Self-organizing and collective behavior techniques, with their "fire and forget" nature, may lead system managers to be more complacent in their security oversight. A resilient self-organizing system that effectively manages itself is far less likely to be overseen with a watchful eye than a system prone to periodic breakdowns and requiring episodic maintenance. If system administrators fall prey to this temptation, a wily attacker could take advantage of their lack of vigilance.

## PERSONNEL MANAGEMENT

### Vulnerabilities Addressed (Directly)

*Difficulty of Management.* Confidence-building activities, such as training of staff, awareness exercises and tests, and design of systems to aid in their effective operation, directly contribute to the ability to manage and operate a complex information system.

*Complacency/Co-optability.* Combating complacency in the operation and maintenance of a system, and its ability to be co-opted by an unauthorized person, should be directly addressed as part of the training and education of the system staff. The human-computer interface for system operation and monitoring should be designed and configured so that unusual patterns of behavior (that might indicate the co-opting of a user account, for example) are quickly recognizable.

### Vulnerabilities Addressed (Indirectly)

*Malleability.* Personnel checks and training may indirectly address malleability by reducing the likelihood that insiders will attempt to exploit this vulnerability and by limiting the impact of such exploitation when it does occur.

*Lack of Self-Awareness.* The vigilance of well-trained, reliable users may provide some added self-awareness to a system.

*Electronic Accessibility.* If authorized users have been thoroughly investigated and are properly trained, then it is less likely that they will abuse their electronic access or inadvertently allow it to be abused by others.

*Physical Accessibility.* Background checks and training will limit the likelihood and impact of attacks that exploit physical access.

## CENTRALIZED MANAGEMENT OF INFORMATION RESOURCES

### Vulnerabilities Addressed (Directly)

*Difficulty of Management.* Centralized management can institute uniform policies and procedures for system management that aid in overall system management and control. As one example, (semi-) automated distribution and installation of bug fixes from a central source might aid in maintaining the security of a distributed system.

### Vulnerabilities Addressed (Indirectly)

*Lack of Recoverability.* Centralized management will have greater resources available, perhaps fungible to some extent among various systems under its control, to aid in quickly recovering portions of the overall system that have become inoperable.

*Lack of Self-Awareness.* Centralized management can institute policies and procedures across various system components, giving the central source greater awareness of attacks—or of abnormal patterns of system behavior—than any individual system component might have.

### Potential Vulnerabilities Incurred (Directly)

*Centralization.* Centralized management, if vital to the operation of a system, itself becomes a centralized system "component" that could be attacked or disabled.

### Potential Vulnerabilities Incurred (Indirectly)

*Homogeneity.* Centralized management may have a tendency to institute policies, for convenience or greater simplicity of operation, that increase system homogeneity.

*Predictability.* Centralized management would tend to institute uniform policies and procedures throughout the sites and nodes of a distributed system. Those uniform procedures increase the predictability of the system's operation, thereby aiding intruders in predicting the effects of their interventions.

*Dependency.* Centralized management necessarily requires more communication, possibly across large distances, than a decentralized approach would involve. Thus, applying techniques of this type could make a system more dependent on external communications infrastructures.

## THREAT/WARNING RESPONSE STRUCTURE

### Vulnerabilities Addressed (Directly)

*Complacency/Co-optability.* Having a graduated threat and warning response structure addresses possible complacency in system operation, since contingencies are planned to allow rapid and somewhat tailored response to situations as they arise.

*Electronic Accessibility.* A graduated response structure would tend to limit electronic accessibility (e.g., by limiting or closing down firewall options) as the threat or warning level increases.

*Transparency.* A graduated response structure would increase the levels of deception used and limit accessibility of system components, making the operation and structure of an information system less transparent to intruders.

*Physical Accessibility.* Graduated response levels would restrict physical accessibility to a system at higher threat or warning levels. For example, more stringent access controls to system sites would be instituted at a high threat level.

### Vulnerabilities Addressed (Indirectly)

*Rigidity.* A graduated system response to levels of threat or warning adds tailored flexibility to system behavior, thereby reducing its rigidity. However, if all canned and planned responses are scripted in advance for various threat levels, the system may remain quite rigid in its behaviors.

*Malleability.* At higher threat or warning levels, a graduated response system would tend to reduce the malleability of a system, since access would be more controlled and more difficult, and there would tend to be less flexibility allowed in modifying system behavior and components.

*Gullibility.* Greater controls and restrictions on system access at higher threat or warning levels would tend to ameliorate any gullibility of the system in overly trusting its users and inputs.

*Capacity Limits.* At higher threat or warning levels, many less-important administrative processes are removed from critical systems, freeing up system capacity for critical processes and data.

*Lack of Recoverability.* A system that has difficulty recovering from a serious attack could be protected with a response structure that would reduce the chances of an attack reaching the point where recovery would be extremely difficult or costly.

*Electromagnetic Susceptibility.* As greater physical security measures and site controls are instituted at higher threat or warning levels, there is some reduction in vulnerability to electromagnetic attacks on systems from local hand-carried or vehicle-based electromagnetic weapons.

*Dependency.* At higher threat or warning levels, measures should have been taken within a graduated response strategy to reduce dependency on such auxiliary systems as power, air-conditioning, water, and so on. For example, auxiliary generators might be placed in a state of readiness or placed online.

### Potential Vulnerabilities Incurred (Indirectly)

*Difficulty of Management.* Implementing a response structure would almost certainly involve new management tasks, such as selecting and adjusting the triggering thresholds for the various response levels, which could make managing the system more difficult and possibly increase the associated vulnerability.

# INFORMATION ASSURANCE RESEARCH PROJECTS EXAMINED

### Table G.1

### Information Assurance Research Projects Examined

| Code | Project Name | Research Institution | Researchers |
|------|--------------|----------------------|-------------|
| T1 | Complete, Automatic Analysis of Cryptographic Protocols | Arca Systems, Inc. | Stephen Brackin |
| T2 | Policy Based Dynamic Security Management | BBN | N/A |
| T3 | Internet Routing Infrastructure Security | BBN Systems and Technologies | Stephen Kent |
| T4 | An Open Implementation Toolkit for Creating Adaptable Distributed Applications | BBN Systems and Technologies, A Division of BBN Corporation | Richard Schantz |
| T5 | Independent Monitoring for Network Survivability | Bell Communications Research, Inc. (Bellcore) | Christian Huitema |
| T6 | Survivable Active Networking | Bell Communications Research, Inc. (Bellcore) | Mark Segal |
| T7 | Intrusion Tolerance via Threshold Cryptography (ITTC) | Bell Communications Research, Inc. (Bellcore) | Dan Boneh |
| T8 | Traffic Management for Survivability of Large-Scale Networks | Bell Communications Research, Inc. (Bellcore) | Jonathan Wang |
| T9 | Adaptive System Security Policies | Boeing Defense & Space Group | Dan Schnackenberg |
| T10 | Dynamic, Cooperating Boundary Controllers | Boeing Defense & Space Group | Dan Schnackenberg |

## Table G.1—continued

| | | | |
|---|---|---|---|
| T11 | Electronic Commerce: The NetBill Project | Carnegie Mellon University | Marvin Sirbu |
| T12 | Reasoning About Implicit Invocation Systems | Carnegie Mellon University | David Garlan |
| T13 | The Fox Project: Advanced Languages for Systems Software | Carnegie Mellon University | Robert Harper |
| T14 | End-to-End Reservation Services in Real-Time Mach | Carnegie Mellon University | Raj Rajkumar |
| T15 | Resource-Centric Microkernel and Communication Services | Carnegie Mellon University | Raj Rajkumar |
| T16 | Concerning Invictus: Detection of Unanticipated and Anomalous Events | Carnegie Mellon University | Roy Maxion |
| T17 | Fraud and Intrusion Detection for Financial Information Systems using Meta-Learning Agents | Columbia University | Salvatore J. Stolfo |
| T18 | MARKETNET: A Survivable, Market-Based Architecture for Large-Scale Information Systems | Columbia University | Yechiam Yemini |
| T19 | Modeling and Testing the MK++ Kernel | Computational Logic, Inc. (CLI) | William Bevier |
| T20 | Software Tools for Enhanced Computer Security | Computer Operations, Audit, and Security Technology Laboratory, Purdue University | Eugene Spafford |
| T21 | Analysis and Response for Intrusion Detection in Large Networks | Computer Science Lab, SRI International | Phillip Porras |
| T22 | Secure Real-Time Multicast: The Ensemble System | Cornell University | Kenneth P. Birman |
| T23 | Foundations and Support for Survivable Systems | Cornell University | Fred Schneider |
| T24 | Integrating Formal and Informal Techniques | Department of Computer Science, Michigan State University | Betty Cheng |
| T25 | Highly Structured Architecture for High Integrity Networks | Department of Computer Science, University of Arizona | Larry Peterson |
| T26a | Development of Real-Time Secure Operating Systems | Hughes Aircraft Company | N/A |

## Table G.1—continued

| | | | |
|---|---|---|---|
| T26b | Development of Real-Time Secure Operating Systems | Intel Corporation | N/A |
| T27 | Composition, Proof, and Reuse (CPR) for Survivable Systems | Kestrel Institute | Allen Goldberg |
| T28 | Adaptable, Dependable Wrappers | Key Software, Inc. | Franklin Webber |
| T29 | Programming-Language Structures for Representing and Optimizing Operating-System Resources | Massachusetts Institute of Technology | Gerald Sussman |
| T30 | Celestial Security Service Management Architecture | MCNC | Fengmin Gong |
| T31 | Key Agile Cryptographic Systems | MCNC | Fengmin Gong |
| T32 | Network Intrusion Detection | MCNC | Y. Frank Jou |
| T33 | Security for Distributed Computer Systems | MIT Laboratory for Computer Science | Shafi Goldwasser |
| T34 | Adaptive Network Security Management | Mountain Wave, Inc. | Juanita Koilpillai |
| T35 | An Environment for Developing Secure Software | Naval Postgraduate School | Dennis Volpano |
| T36 | Onion Routing: Anonymous Communications Infrastructure | Naval Research Laboratory | Paul Syverson |
| T37 | Formal Analysis of Internet Security Protocols | Naval Research Laboratory | Catherine Meadows |
| T38 | Survivability in Object Services Architectures | Object Services and Consulting, Inc. | David Wells |
| T39 | Task-Based Authorization | Odyssey Research Associates | Roshan Thomas |
| T40 | Computational Immunology for Distributed Large Scale Systems | Odyssey Research Associates | Maureen Stillman |
| T41 | Security Engineering for High-Assurance, Policy-Based Applications | Odyssey Research Associates | Richard Platek |
| T42 | Immunix Project | Oregon Graduate Institute for Science & Technology | Calton Pu |
| T43 | The Heterodyne Project | Oregon Graduate Institute for Science & Technology | David Maier |
| T44 | Secure Mobile Networking | Portland State University | John McHugh |

**Table G.1—continued**

| T45 | Enhanced Intrusion and Misuse Detection Technology | Purdue University | Eugene Spafford |
|-----|-----|-----|-----|
| T46 | Minimum Essential Information Infrastructure | RAND | Robert Anderson |
| T47 | Quantifying Minimum-Time-to-Intrusion Based on Dynamic Software Safety Assessment | Reliable Software Technologies Corporation | Jeffrey Voas |
| T48 | Dynamic Security Analysis of COTS Applications | Reliable Software Technologies Corporation | Anup Ghosh |
| T49 | Software Mutation for Survivability | Reliable Software Technologies Corporation | Christoph Michael |
| T50 | Secure Heterogeneous Application Runtime Environment (SHARE) | Sanders, A Lockheed Martin Company | Jeff Smith |
| T51 | Puma-Based, Real-Time, Secure OS | Sandia National Laboratories | David Greenberg |
| T52 | Kernel Hypervisors | Secure Computing Corporation | Dick O'Brien |
| T53 | Composability for Secure Systems | Secure Computing Corporation | Duane Olawsky |
| T54 | MAUDE: A Wide-Spectrum Formal Language for Secure Active Networks | SRI International | Jose Meseguer |
| T55 | Secure Access Wrapper | SRI International | Steven Dawson |
| T56 | SDTP: An Open Standard for Secure Distribution Transaction Processing | SRI International | Robert Riemenschneider |
| T57 | Explaining and Recovering from Computer Break-ins | SRI International | Douglas Moran |
| T58 | Survivable Loosely Coupled Architectures | SRI International, Computer Science Laboratory | John Rushby |
| T59 | Formally Verified Hardware Encapsulation for Security and Safety | SRI International, Computer Science Laboratory | John Rushby |
| T60 | Highly Assured and Fault-Tolerant Security in Distributed Systems | SRI International, Computer Science Laboratory | John Rushby |
| T61 | Semantic Interoperation of Open Systems | Stanford University | Carolyn Talcott |

## Table G.1—continued

| | | | |
|---|---|---|---|
| T62 | Fault Handling and Customization | The Open Group Research Institute of the Open Software Foundation, Inc. | Paul Dale |
| T63 | Adage: Authorization for Distributed Applications and Groups | The Open Group Research Institute of the Open Software Foundation, Inc. | Mary Ellen Zurko |
| T64 | MANET: Mobile Agents for Network Trust | The Open Group Research Institute of the Open Software Foundation, Inc. | David Black |
| T65 | Protocols for Secure and Survivable Active Internetworking | The Regents of the University of California, Santa Cruz (UCSC) | J. J. Garcia-Luan-Aceves |
| T66 | Enabling Real-Time Fault-Tolerant Applications | The University of Michigan | Kang Shin |
| T67 | Internet Infrastructure Protection | Trusted Information Systems | Russ Mundy |
| T68 | Policy-Based Cryptographic Key Release System | Trusted Information Systems | Dennis Branstad |
| T69 | Secure Active Network Prototypes | Trusted Information Systems | Russ Mundy |
| T70 | Advanced Security Proxy Technology for High-Confidence Networks | Trusted Information Systems | E. John Sebes |
| T71 | Dynamic Cryptographic Context Management | Trusted Information Systems | Dennis Branstad |
| T72 | Internet Safety and Security Task: Internet Safety Through Type-Enforcing Firewalls | Trusted Information Systems | Martha Branstad |
| T73 | User-Level Truffles | Trusted Information Systems | Martha Branstad |
| T74 | Generic Software Wrappers for Security and Reliability | Trusted Information Systems | Martha Branstad |
| T75 | Composable Replaceable Security Services | Trusted Information Systems | Richard Feiertag |
| T76 | SIGMA: Security and Interoperability for Heterogeneous Distributed Systems | Trusted Information Systems | Terry Benzel |
| T77 | INFOSEC for Networked Systems Task: Security Consulting and Cooperative Research | Trusted Information Systems | Martha Branstad |

## Table G.1—continued

| T78 | Extensible OS Security | Trusted Information Systems | Dennis Hollingsworth |
|---|---|---|---|
| T79 | Secure Virtual Enclaves | Trusted Information Systems | E. John Sebes |
| T80 | International Cryptography Experiment (ICE) | Trusted Information Systems | David Balenson |
| T81 | Security Agility for Dynamic Execution Environments | Trusted Information Systems | Lee Badger |
| T82 | Infosec for Networked Computers | Trusted Information Systems | N/A |
| T83 | Internet Security Technology | U.S. Naval Research Laboratory | Ron Lee |
| T84 | Cactus: An Integrated Framework for Dynamic Fine-Grain QoS | University of Arizona | Richard D. Schlichting |
| T85 | Adaptive Distributed Systems | University of Arizona | Richard D. Schlichting |
| T86 | Secure Execution of Mobile Programs | University of California, Davis | Raju Pandey |
| T87 | Transparent Virtual Mobile Environment (Traveler) | University of California, Los Angeles | Leonard Kleinrock |
| T88 | The Design of Fault-Tolerant Real-Time Systems | University of California, Santa Barbara | Michael Melliar-Smith |
| T89 | The Eternal System | University of California, Santa Barbara | Louise Moser |
| T90 | Secure Multicast Protocols for Group Communication | University of California, Santa Barbara | Louise Moser |
| T91 | NetSTAT: A Model-Based Real-Time Intrusion Detection System | University of California, Santa Barbara | Richard Kemmerer |
| T92 | An Agent-Based Architecture for Supporting Application Aware Security | University of Illinois at Urbana-Champaign | Roy Campbell |
| T93 | Enhancing Survivability with Distributed Adaptive Coordination | University of Massachusetts at Amherst | Victor Lesser |
| T94 | Market-Based Adaptive Architectures for Information Survivability | University of Michigan | Michael Wellman |
| T95 | Self-Configuring Survivable Multi-Networks | University of Missouri, Kansas City | Deep Medhi |
| T96 | CLIQUES: Security Services for Group Communication | University of Southern California, Information Sciences Institute | Herbert Schorr |

## Table G.1—continued

| T97 | SILDS: Security Infrastructure for Large Distributed Systems | University of Southern California, Information Sciences Institute | Clifford Neuman |
|---|---|---|---|
| T99 | Advanced Scalable Network Technology | University of Southern California, Information Sciences Institute | John Granacki |
| T100 | Intrusion Tolerance for Legacy Applications | University of Texas at Austin | Aleta Ricciardi |
| T101 | Mach 4 Kernel and IDL Infrastructure for Security | University of Utah | N/A |
| T102 | Survivability Architectures | University of Virginia | David Notkin |
| T103 | Critical Analysis of the Use of Redundancy to Achieve Survivability in the Presence of Malicious Attacks | University of Wisconsin, Milwaukee | Yvo Desmedt |
| T104 | CRISIS: Critical Resource Allocation and Intrusion Response for Survivable Information Systems | University of Southern California Information Sciences Institute | Herb Schorr |
| S1 | Information Assurance Technologies for the NGII | Secure Computing Corp. | Dick O'Brien |
| S2 | A Security and Protection Foundation for the NGII Reference Architecture | Trusted Information Systems | Dan Sterne |
| S3 | Access Control and Security Management for the NGII | The Open Group Research Institute of the Open Software Foundation, Inc. | Mary Ellen Zurko |
| S4 | A Security Management Foundation for the NGII | Trusted Information Systems | Richard Feiertag |
| S5 | NETWORK RADAR: Surveillance and Tracking in Computer Networks | Net Squared | Todd & Antoinette Heberlein |
| S6 | Automatic Response to Intrusion | Boeing Defense & Space Group | D. Schnackenberg |
| B1 | Self signature generated by system calls | University of New Mexico, Santa Fe Institute | S. Forrest, S. Hofmeyr, A. Somayaji |
| B2 | Lymphocyte interrogation and punishment of processes | University of New Mexico, Santa Fe Institute | S. Forrest, S. Hofmeyr, A. Somayaji |
| B3 | Lymphocytes with multiple detectors and selective replication | University of New Mexico, Santa Fe Institute; IBM Watson Research Center | S. Forrest, S. Hofmeyr, A. Somayaji, J. Kephart et al. |

**Table G.1—continued**

| B4 | Lymphocytes with selectively lengthened lifespans | University of New Mexico, Santa Fe Institute | S. Forrest, S. Hofmeyr, A. Somayaji |
|---|---|---|---|
| B5 | Lymphocyte migration between trusted computers | University of New Mexico, Santa Fe Institute; IBM Watson Research Center | S. Forrest, S. Hofmeyr, A. Somayaji, J. Kephart et al. |
| B6 | Dedicated computer thymus to produce lymphocytes | University of New Mexico, Santa Fe Institute | S. Forrest, S. Hofmeyr, A. Somayaji |
| B7 | Dedicated computers acting as immune organs | University of New Mexico, Santa Fe Institute | S. Forrest, S. Hofmeyr, A. Somayaji |
| B8 | Propagation of "kill signal" to contain damage | IBM Watson Research Center | J. Kephart et al. |
| B9 | Use of intrusion detection as an acquired immune response with information migration between machines | University of New Mexico, Santa Fe Institute; IBM Watson Research Center | S. Forrest, S. Hofmeyr, A. Somayaji, J. Kephart et al. |
| B10 | Self-organization and emergent properties in autonomous agent collections | University of New Mexico, Santa Fe Institute; Ecole Nationale Superieure des Telecommunications de Paris (ENST) | S. Forrest, S. Hofmeyr, A. Somayaji, E. Bonabeau et al. |
| B11 | Mobile software agents for decentralized decisionmaking and optimization | University of New Mexico, Santa Fe Institute; Ecole Nationale Superieure des Telecommunications de Paris (ENST) | S. Forrest, S. Hofmeyr, A. Somayaji, E. Bonabeau et al. |
| B12 | Collective behaviors using autonomous agents | COAST Laboratories (at Purdue University) | E. Spafford, M. Crosbie |
| B13 | Self-healing in networks using autonomous agents | MCI, British Telecom | N/A |
| B14 | Load-balancing in telecommunications networks | University of New Mexico, Santa Fe Institute; Ecole Nationale Superieure des Telecommunications de Paris (ENST) | S. Forrest, S. Hofmeyr, A. Somayaji, E. Bonabeau et al. |
| B15 | Randomized compilation techniques to create diversity | University of New Mexico, Santa Fe Institute | S. Forrest, S. Hofmeyr, A. Somayaji |

### Table G.1—continued

| B16 | Systematic specialization to complicate execution paths | Oregon Graduate Institute for Science & Technology | Calton Pu |
|-----|----------------------------------------------------------|-----------------------------------------------------|-----------|
| B17 | Epidemiological study of platform/component/ software diversity | IBM Watson Research Center | J. Kephart et al. |
| B18 | Decoy-infection routines to capture viruses | IBM Watson Research Center | J. Kephart et al. |
| B19 | Software code optimization by genetic programming techniques | University of New Mexico, Santa Fe Institute | S. Forrest, S. Hofmeyr, A. Somayaji |
| B20 | Imperfect or approximate detection algorithms | University of New Mexico, Santa Fe Institute; Stanford Research Institute | S. Forrest, S. Hofmeyr, A. Somayaji |

NOTES: This list does not include the NSA projects. Information regarding these projects was made available to us and used in our classification of them in Chapter Four, but the details of specific projects are not publicly releasable. The letter in each project number indicates to which group it belongs: T = DARPA ITO; S = DARPA ISO; B = biomimetic research. Project T98 was not included because it was removed from the publicly available list of DARPA ITO projects.

# REFERENCES

Appleby, S., and S. Steward, "Mobile Software Agents for Control in Telecommunications Networks" *British Telecommunications Technology Journal,* Vol. 12, 1994, pp. 104–113.

Arquilla, John, "The Great Cyberwar of 2002," *Wired,* February, 1998.

Arquilla, John, and David Ronfeldt, eds., *In Athena's Camp: Preparing for Conflict in the Information Age,* Santa Monica, Calif.: RAND, MR-880-OSD/RC, 1997.

Bertsekas, Dimitri, and Robert Gallager, *Data Networks,* 2nd edition, Englewood Cliffs, N.J.: Prentice Hall, 1992.

Bonabeau E., F. Henaux, S. Guerin, D. Snyers, P. Kuntz, and G. Theraulaz, *Routing in Telecommunications Networks With "Smart" Ant-Like Agents,* Santa Fe Institute & Ecole Nationale Superieure des Telecommunications de Paris, 1998.

Cheswick, W. R., and S. M. Bellovin, *Firewalls and Internet Security: Repelling the Wily Hacker,* Reading, Mass.: Addison-Wesley, 1994.

Cohen, F., *The Deception ToolKit,* initial release, available at http://www.all.net/, 1998.

Courand, G., *Counter Deception,* Office of Naval Research, 1989.

Crosbie, M., and G. Spafford, *Defending a Computer Using Autonomous Agents,* Lafayette, Ind.: COAST Labs, Purdue University, 1995.

Crosbie, M., and G. Spafford, *Applying Genetic Programming to Intrusion Detection*, Lafayette, Ind.: COAST Labs, Purdue University, 1995.

Cruickshank, C., *Deception in World War Two*, London: Oxford University Press, 1979.

Dam, Kenneth W., and Herbert S. Lin, *Cryptography's Role in Securing the Information Society*, Washington, D.C.: National Academy Press, 1996.

Defense Advanced Research Projects Agency (DARPA), *Information Assurance Security Focus Group (1997) Security Architecture for the AITS Reference Architecture, Revision 1.0*, December 22, 1997. Also available at http://www.darpa.mil/iso/ia/, 1997.

Defense Information Systems Agency (DISA), *Defense Information Infrastructure (DII) Common Operating Environment (COE) Integration and Runtime Specification (I&RTS)*, Version 2.0 (preliminary), October 23, 1995.

Department of Defense, *Critical Asset Assurance Program*, DoD Directive 5160.54, January 20, 1998 (available at http://web7.whs.osd.mil/corres.htm/).

Defense Science Board (DSB), *Report of the Defense Science Board Task Force on Information Warfare—Defense (IW-D)*, Washington, D.C.: Office of the Under Secretary of Defense for Acquisition and Technology, November 1996.

Dewar, M., *The Art of Deception in Warfare*, United Kingdom: David & Charles, 1989.

Dunnigan, J., and A. Nofi, *Victory and Deceit*, New York: William Morrow & Co., 1996.

Everest Consulting, *Deception: Fact and Folklore*, McLean, Va.: Central Intelligence Agency, 1980.

Farnham, David E., "Logic for Intelligence Analysts," in Ronald Garst, ed., *A Handbook of Intelligence Analysis*, Defense Intelligence College, 1988.

Feer, Frederic S., unpublished RAND research on a survey and analysis of technology and military deception.

Feldman, Phillip M., unpublished RAND research on potential implications for the Air Force of the vulnerabilities of the public switched network.

Forrest, S., A. Perelson, L. Allen, and R. Cherukuri, "Self-Nonself Discrimination in a Computer," *Proceedings of the 1994 IEEE Computer Society Symposium on Research in Security and Privacy,* 1994.

Forrest S., A. Somayaji, and D. H. Ackley, "Building Diverse Computer Systems," in *Sixth Workshop on Hot Topics in Operating Systems,* 1997.

Howard, J. D., *An Analysis of Security Incidents on the Internet, 1989–1995,* (Ph.D. thesis in Engineering and Public Policy), Pittsburgh, Pa.: Carnegie-Mellon University, 1997.

Hundley, Richard O., and Robert H. Anderson, "Emerging Challenge: Security and Safety in Cyberspace," *IEEE Technology and Society Magazine,* Vol. 14, No. 4, Winter 1995–96. Also available as RAND RP-484, and as Chapter 10 of Arquilla and Ronfeldt, 1997.

Kahneman, D., and A. Tversky, "Choices, Values, Frames," *American Psychologist,* Vol. 39, No. 4, 1983, pp. 341–350.

Kephart, J. O., "A Biologically Inspired Immune System for Computers," in R. A. Brooks and P. Maes, eds., *Artificial Life IV: Proceedings of the Fourth International Workshop on the Synthesis and Simulation of Living Systems,* Cambridge, Mass.: MIT Press, pp. 130–139, 1994a.

Kephart, J. O., "How Topology Affects Population Dynamics," in C. Langton, ed., *Artificial Life III: Studies in the Sciences of Complexity,* Cambridge, Mass.: MIT Press, pp. 447–463, 1994b.

Kephart, J. O., "Biologically Inspired Defenses Against Computer Viruses," *Proceedings of International Joint Conference on Artificial Intelligence (IJCAI '95),* Montreal, pp. 985–996, August 19–25, 1995.

Kephart, J. O., Gregory B. Sorkin, Morton Swimmer, and Steve R. White, *Blueprint for a Computer Immune System,* New York: IBM Thomas J. Watson Research Center, 1997. (Originally presented at the Virus Bulletin International Conference in San Francisco, Calif., October 1–3, 1997.)

Kephart J. O., S. R. White, and D. M. Chess, "Computers and Epidemiology," *IEEE Spectrum,* May 1993, pp. 20–26.

Khalilzad, Zalmay M., and John White, eds., *Strategic Appraisal: The Changing Role of Information in Warfare,* Santa Monica, Calif.: RAND, MR-1016-AF, forthcoming.

Lambert, D. R., *A Cognitive Model for the Exposition of Human Deception and Counterdeception,* Washington, D.C.: Naval Materiel Command, NOSC TR 1076, 1987.

Lloyd, M., *The Art of Military Deception,* London: Leo Cooper, 1997.

McCleskey, E., *Applying Deception to Special Operations Direct Action Missions,* Ft. Belvoir, Va.: Defense Technical Information Center, Defense Intelligence College, 1991.

Molander, Roger C., Andrew S. Riddile, and Peter A. Wilson, *Strategic Information Warfare: A New Face of War,* Santa Monica, Calif.: RAND, MR-661-OSD, 1996.

Molander, Roger C., Peter A. Wilson, David A. Mussington, and Richard Mesic, *Strategic Information Warfare Rising,* Santa Monica, Calif.: RAND, MR-964-OSD, 1998.

Molander, Roger C., and Peter A. Wilson, *The Day After . . . in the American Strategic Infrastructure,* Santa Monica, Calif.: RAND, MR-963-OSD, 1998.

Neumann, P. G., *Computer Related Risks,* New York: Addison-Wesley, 1995.

Perrow, Charles, *Normal Accidents: Living with High-Risk Technologies,* New York: Basic Books, 1984.

President's Commission on Critical Infrastructure Protection (PCCIP), *Critical Foundations: Protecting America's Infrastruc-*

*tures: The Report of the President's Commission on Critical Infrastructure Protection,* available at http://www.pccip.gov/, 1997.

Schneider, Fred B., ed., *Trust in Cyberspace,* Washington D.C.: National Academy Press, 1998.

Schneier, Bruce, *Applied Cryptography,* 2nd edition, New York: John Wiley & Sons, 1996.

Schoonderwoerd, R., *Collective Intelligence for Network Control,* the Netherlands: Delft University of Technology, 1996.

Schoonderwoerd, R., O. Holland, J. Bruten, and L. Rothkrantz, "Ant-Based Load Balancing in Telecommunications Networks," *Adaptive Behavior,* Vol. 5, 1997, pp. 169–207.

Somayaji, A., S. Hofmeyr, and S. Forrest, "Principles of a Computer Immune System," *Proceedings, New Security Paradigms Workshop '97,* forthcoming.

Tversky, A., and D. Kahneman, "Judgment Under Uncertainty: Heuristics and Biases," *Science,* Vol. 185, 1974, p. 4157.

U.S. Army, *Field Manual 90-2: Battlefield Deception,* 1988.

U.S. Joint Chiefs of Staff, *Joint Pub 3-58: Joint Doctrine for Military Deception,* 1994.

Whaley, B., *Stratagem: Deception and Surprise in War,* Cambridge, Mass.: Center for International Studies, 1969.

Whaley, B., "Toward a General Theory of Deception," in John Gooch and Amos Perlmutter, eds., *Military Deception and Strategic Surprise,* London: Frank Cass, 1982.

Whelan, William J., and John Arbeeny, unpublished RAND research on operations flexibility and deception.

Woodward, John, "Information Assurance: The Key Is the Right Combination," *C4I News,* December 4, 1997.

Wright, S., "General Network Scenario: Network Evolution," briefing slides from TranSwitch seminar on *Voice over ATM,* June 25, 1996.

Zorpette, G., "Special Report: Keeping the Phone Lines Open," *IEEE Spectrum*, June 1989.